LECTURE NOTES ON
PRINCIPLES OF
PLASMA PROCESSING

LECTURE NOTES ON PRINCIPLES OF PLASMA PROCESSING

Francis F. Chen

Electrical Engineering Department

and

Jane P. Chang

Chemical Engineering Department

University of California
Los Angeles, California

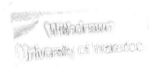

Kluwer Academic / Plenum Publishers
New York, Boston, Dordrecht, London, Moscow

Library of Congress Cataloging-in-Publication Data

Chen, Francis F., 1929–
      Lecture notes on principles of plasma processing/Francis F. Chen, Jane P. Chang.
         p.   cm.
      Includes bibliographical references and index.
      ISBN 0-306-47497-2
         1. Plasma engineering.   2. Plasma chemistry.   3. Plasma dynamics.   I. Title: Principles of
plasma processing.   II. Chang, Jane P., 1970–   III. Title.

TA2020 .C454 2003
621.044—dc21

                                                                                                          2002040665

ISBN: 0-306-47497-2

©2003 Kluwer Academic / Plenum Publishers, New York
233 Spring Street, New York, New York 10013

http://www.wkap.nl/

10   9   8   7   6   5   4   3   2   1

Printed in the United States of America

PRINCIPLES OF PLASMA PROCESSING

PREFACE

Reference books used in this course

We want to make clear at the outset what this book is NOT. It is *not* a polished, comprehensive textbook on plasma processing, such as that by Lieberman and Lichtenberg. Rather, it is an informal set of lecture notes written for a nine-week course offered every two years at UCLA. It is intended for seniors and graduate students, especially chemical engineers, who have had no previous exposure to plasma physics. A broad range of topics is covered, but only a few can be discussed in enough depth to give students a glimpse of forefront research. Since plasmas seem strange to most chemical engineers, plasma concepts are introduced as painlessly as possible. Detailed proofs are omitted, and only the essential elements of plasma physics are given. One of these is the concept of sheaths and quasineutrality. Sheaths are dominant in plasma "reactors," and it is important to develop a physical feel for their behavior.

Good textbooks do exist. Two of these, to which we make page references in these notes for those who want to dig deeper, are the following:

M.A. Lieberman and A.J. Lichtenberg, *Principles of Plasma Discharges and Materials Processing* (John Wiley, New York, 1994).

F.F. Chen, *Introduction to Plasma Physics and Controlled Fusion*, Vol. 1, 2nd ed. (Plenum Press, 1984).

In addition, more topics and more detail are available in unpublished notes from short courses offered by the American Vacuum Society or the Symposium on Plasma and Process Induced Damage. Lecture notes by such specialists as Prof. H.H. Sawin of M.I.T. are more comprehensive. Our aim here is to be comprehensible..

The lectures on plasma physics (Part A) and on plasma chemistry (Part B) are interleaved in class meetings but for convenience are printed consecutively here, since they were written by different authors. We have tried to keep the notation the same, though physicists and chemists do tend to express the same formula in different ways. There are no doubt a few mistakes; after all, these are just notes. As for the diagrams, we have given the source wherever possible. Some have been handed down from antiquity. If any of these are yours, please let us know, and we will be glad to give due credit. The diagrams are rather small in printed form. The CD which

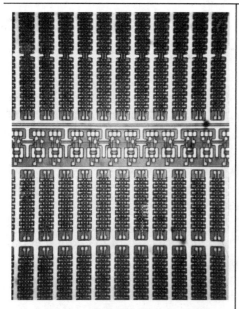

A small section of a memory chip.

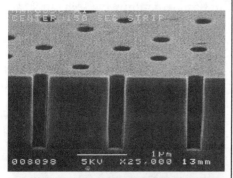

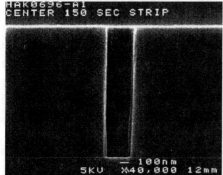

Straight holes like these can be etched
only with plasmas

accompanies the text has color figures that can be expanded for viewing on a computer monitor. There are also sample homework problems and exam questions there.

Why study plasma processing? Because we can't get along without computer chips and mobile phones these days. About half the steps in making a semiconductor circuit require a plasma, and plasma machines account for most of the equipment cost in a "fab." Designers, engineers, and technicians need to know how a plasma behaves. These machines have to be absolutely reliable, because many millions of transistors have to be etched properly on each chip. It is amazing that this can be done at all; improvements will certainly require more plasma expertise. High-temperature plasmas have been studied for decades in connection with controlled fusion; that is, the production of electric power by creating miniature suns on the earth. The low-temperature plasmas used in manufacturing are more complicated because they are not fully ionized; there are neutral atoms and many collisions. For many years, plasma sources were developed by trial and error, there being little understanding of how these devices worked. With the vast store of knowledge built up by the fusion effort, the situation is changing. Partially ionized, radiofrequency plasmas are being better understood, particularly with the use of computer simulation. Low-temperature plasma physics is becoming a real science. This is the new frontier. We hope you will join in the exploration of it.

Francis F. Chen
Jane P. Chang
Los Angeles, 2002

TABLE OF CONTENTS

Lecture Notes on
PRINCIPLES OF
PLASMA PROCESSING

Diagrams can be enlarged on a computer by using the CD-ROM.

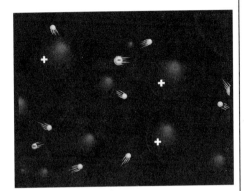

Ions and electrons make a plasma

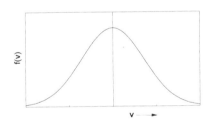

A Maxwellian distribution

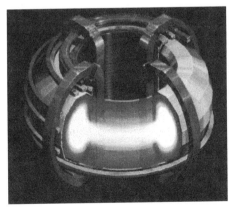

A "hot" plasma in a fusion reactor

PRINCIPLES OF PLASMA PROCESSING
Course Notes: Prof. F.F. Chen

PART A1: INTRODUCTION TO PLASMA SCIENCE

I. WHAT IS A PLASMA?

Plasma is matter heated beyond its gaseous state, heated to a temperature so high that atoms are stripped of at least one electron in their outer shells, so that what remains are positive ions in a sea of free electrons. This *ionization* process is something we shall study in more detail. Not all the atoms have to be ionized: the cooler plasmas used in plasma processing are only 1-10% ionized, with the rest of the gas remaining as neutral atoms or molecules. At higher temperatures, such as those in nuclear fusion research, plasmas become fully ionized, meaning that all the particles are charged, not that the nuclei have been stripped of all their electrons.

We can call a plasma "hot" or "cold", but these terms have to be explained carefully. Ordinary fluids are in thermal equilibrium, meaning that the atoms or molecules have a Maxwellian (Gaussian) velocity distribution like this:

$$f(v) = Ae^{-(\frac{1}{2}mv^2/KT)},$$

where A is a normalization factor, and K is Boltzmann's constant. The parameter T, then, is the temperature, which determines the width of the distribution. In a plasma, the different species—ions, electrons, and neutrals—may have different temperatures: T_i, T_e, and T_n. These three (or more, if there are different kinds of ions or atoms) interpenetrating fluids can move through one another, but they may not collide often enough to equalize the temperatures, because the densities are usually much lower than for a gas at atmospheric pressure. However, each species usually collides with itself often enough to have a Maxwellian distribution. Very hot plasmas may be non-Maxwellian and would have to be treated by "kinetic theory".

A "cool" plasma would have to have an electron temperature of at least about 10,000°K. Then the fast electrons in the "tail" of the distribution would be energetic enough to ionize atoms they collide with often enough to overcome recombination of ions and electrons back into neutrals. Because of the large numbers, it is more convenient to express temperature in electron-volts (eV). When T is such that the energy KT is equal to the

A cooler plasma: the Aurora Borealis

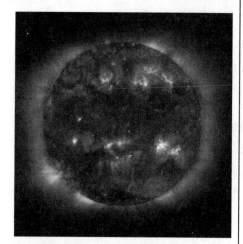

Most of the sun is in a plasma state, especially the corona.

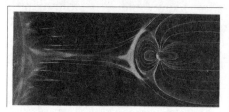

The earth plows through the magnetized interplanetary plasma created by the solar wind.

Comet tails are dusty plasmas.

energy an electron gets when it falls through an electric potential of 1 volt, then we say that the temperature is 1 eV. Note that the average energy of a Maxwellian distribution is $(3/2)KT$, so a 1-eV plasma has average energy 1.5 eV per particle. The conversion factor between degrees and eV is

$$1\,\mathrm{eV} = 11{,}600°\,\mathrm{K}$$

Fluorescent lights contain plasmas with $T_e \approx 1\text{–}2$ eV. Aside from these we do not often encounter plasmas in everyday life, because the plasma state is not compatible with human life. Outside the earth in the ionosphere or outer space, however, almost everything is in the plasma state. In fact, what we see in the sky is visible only because plasmas emit light. Thus, the most obvious application of plasma science is in space science and astrophysics. Here are some examples:

- Aurora borealis
- Solar wind
- Magnetospheres of earth and Jupiter
- Solar corona and sunspots
- Comet tails
- Gaseous nebulae
- Stellar interiors and atmospheres
- Galactic arms
- Quasars, pulsars, novas, and black holes

Plasma science began in the 1920s with experiments on gas discharges by such famous people as Irving Langmuir. During World War II, plasma physicists were called upon to invent microwave tubes to generate radar. Plasma physics got it greatest impetus with the start of research on controlled nuclear fusion in the 1950s. The task is to reproduce on earth the thermonuclear reactions used by stars to generate their energy. This can be done only by containing a plasma of over 10^4 eV (10^8 K). If this enterprise is successful, some say that it will be the greatest achievement of man since the invention of fire, because it will provide our civilization with an infinite source of energy, using only the heavy hydrogen that exists naturally in our oceans.

Another use of plasmas is in generation of coherent radiation: microwave tubes, gas lasers, free-electron lasers, etc. Plasma-based particle accelerators are being developed for high energy physics. Intense X-ray

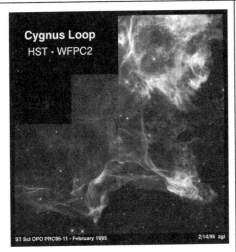

Gaseous nebulae are plasmas.

Plasmas at the birth of stars

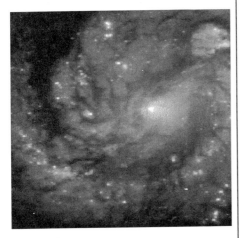

Spiral galaxies are plasmas

sources using *pulsed power* technology simulate nuclear weapons effects. The National Ignition Facility is being built at Livermore *for inertial confinement fusion*. Femtosecond lasers can produce plasmas with such fast rise times that very short chemical and biological events can now be studied. Industrial plasmas, which are cooler, higher pressure, and more complex than those in the applications listed above, are being used for hardening metals, such as airplane turbine blades and automobile parts, for treating plastics for paint adhesion and reduced permeation, for nitriding surfaces against corrosion and abrasion, for forming diamond coatings, and for many other purposes. However, the application of plasma science that is increasingly affecting our everyday life is that of semiconductor production. No fast computer chip can be made without plasma processing, and the industry has a large deficit of personnel trained in plasma science.

II. PLASMA FUNDAMENTALS

Plasma physics has a reputation of being very difficult to understand, and this is probably true when compared with fluid dynamics or electromagnetics in dielectric media. The reason is twofold. First, being a charged fluid, a plasma's particles interact with one another not just by collisions, but by long-range electric and magnetic fields. This is more complicated than treating the charged particles one at a time, such as in an electron beam, because the fields are modified by the plasma itself, and plasma particles can move to shield one another from imposed electric fields. Second, most plasmas are too tenuous and hot to be considered continuous fluids, such as water ($\approx 3 \times 10^{22}$ cm^{-3}) or air ($\approx 3 \times 10^{19}$ cm^{-3}). With particle densities of $10^{9\text{-}13}$ cm^{-3}, plasmas do not always behave like continuous fluids. The discrete nature of the ions and electrons makes a difference; this kind of detail is treated in the *kinetic theory* of plasmas. Fortunately, with a few exceptions, the fluid theory of plasmas is all that is required to understand the behavior of low-temperature industrial plasmas, and the quantum mechanical effects of semiconducting solids also do not come into play.

1. Quasineutrality and Debye length

Plasmas are charged fluids (interpenetrating fluids of ions and electrons) which obey Maxwell's equations, but in a complex way. The electric and magnetic fields in the plasma control the particle orbits. At the same time, the motions of the charged particles can form charge bunches, which create electric fields, or currents,

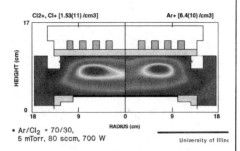

Plasma in a processing reactor (computer model, by M. Kushner)

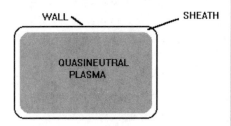

A sheath separates a plasma from walls and large objects.

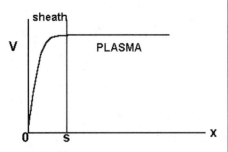

The plasma potential varies slowly in the plasma but rapidly in the sheath.

which create magnetic fields. Thus, the particle motions and the electromagnetic fields have to be solved for in a self-consistent way. One of Maxwell's equations is Poisson's equation:

$$\nabla \cdot \mathbf{D} = \nabla \cdot \varepsilon \mathbf{E} = e(n_i - n_e). \tag{1}$$

Normally, we use ε_0 for ε, since the dielectric charges are explicitly expressed on the right-hand side. For electrostatic fields, E can be derived from a potential V:

$$\mathbf{E} = -\nabla V, \tag{2}$$

whereupon Eq. (1) becomes

$$\nabla^2 V = (e / \varepsilon_0)(n_e - n_i). \tag{3}$$

This equation has a natural scale length for V to vary. To see this, let us replace ∇^2 with $1/L^2$, where L is the length over which V varies. The ratio of the potential energy $|eV|$ of an electron in the electric field to its thermal energy KT_e is then approximately

$$\left| \frac{eV}{KT_e} \right| = L^2 \frac{(n_e - n_i)e^2}{\varepsilon_0 KT_e}. \tag{4}$$

The natural length scale on the right, called the *Debye length*, is defined by

$$\lambda_D = \left(\frac{\varepsilon_0 KT_e}{n_e e^2} \right)^{1/2} \tag{5}$$

In terms of λ_D, Eq. (4) becomes

$$\left| \frac{eV}{KT_e} \right| = \frac{L^2}{\lambda_D^2} \left(1 - \frac{n_i}{n_e} \right). \tag{6}$$

The left-hand side of this equation cannot be much larger than 1, because if a large potential is imposed inside the plasma, such as with a wire connected to a battery, a cloud of charge will immediately build up around the wire to shield out the potential disturbance. When the values of ε_0 and e are inserted, Eq. (5) has the value

$$\lambda_D = 7.4 \sqrt{ \frac{T_e(eV)}{n_e(10^{18}\,\mathrm{m}^{-3})} } \quad \mu\mathrm{m} \tag{7}$$

Thus, λ_D is of order 50 μm for $KT_e = 4$ eV and $n_e = 10^{17}$ m^{-3} or 10^{11} cm^{-3}, a value on the high side for industrial plasmas and on the low side for fusion plasmas. In the main body of the plasma, V would vary over a distance depending on the size of the plasma. If we take L to be of order 10 cm, an average dimension for a laboratory plasma, the factor $(L/\lambda_D)^2$ is of order 10^8, so that n_i must be equal to n_e within one part in 10^8 to keep the LHS reasonably small. In the interior of a plasma, then, the charge densities must be very nearly equal, and we may define a common density, called the *plasma density n*, to be either n_i or n_e. However, there are regions, called *sheaths*, where L is the order of λ_D; there, the ratio n_i / n_e does not have to be near unity.

The condition $n_i \approx n_e$ is called *quasineutrality* and is probably the most important characteristic of a plasma. Charged particles will always find a way to move to shield out large potentials and maintain equal densities of the positive and negative species. We have implicitly assumed that the ions are singly charged. If the ions have a charge Z, the condition of quasineutrality is simply $n_i = Zn_e$. Note that one hardly ever has a whole cubic meter of plasma, at least on the earth; consequently densities are often expressed in cm^{-3} rather than the MKS unit m^{-3}.

If L is of the order of the Debye length, then Eq. (6) tells us that the quasineutrality condition can be violated. This is what happens near the walls around a plasma and near objects, such as probes, inserted into the plasma. Adjacent to the surface, a sheath of thickness about $5\lambda_D$, forms, in which the ions outnumber the electrons, and a strong electric field is created. The potential of the wall is negative relative to the plasma, so that electrons are repelled by a Coulomb barrier. This is necessary because electrons move much faster than ions and would escape from the plasma and leave it positively charged (rather than quasineutral) unless they were repelled by this "sheath drop". We see from Eq. (3) that $V(\mathbf{r})$ would have the right curvature only if $n_i > n_e$; that is, if the sheath is ion-rich. Thus the *plasma potential* tends to be positive relative to the walls or to any electrically isolated object, such as a large piece of dust or a floating probe. Sheaths are important in industrial plasmas, and we shall study them in more detail later.

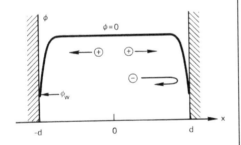

Sheaths form electric barriers for electrons, reflecting most of them so that they escape at the same rate as the slower ions, keeping the plasma quasineutral.

2. Plasma frequency and acoustic velocity

Waves are small, repetitive motions in a continuous medium. In air, we are accustomed to having sound waves and electromagnetic (radio) waves. In water, we have sound waves and, well, water waves. In a plasma, we have electromagnetic waves and two kinds of sound waves, one for each charge species. Of course, if the plasma is partially ionized, the neutrals can have their own sound waves. The sound waves in the electron fluid are called *plasma waves* or *plasma oscillations*. These have a very high characteristic frequency, usually in the microwave range. Imagine that a bunch of electrons are displaced from their normal positions. They will leave behind a bunch of positively charged ions, which will draw the electrons back. In the absence of collisions, the electrons will move back, overshoot their initial positions, and continue to oscillate back and forth. This motion is so fast that the ions cannot move on that time scale and can be considered stationary. The oscillation frequency, denoted by ω_p, is given by

$$\omega_p \equiv \left(\frac{ne^2}{\varepsilon_0 m} \right)^{1/2} \text{rad / sec} \qquad (8)$$

In frequency units, this gives approximately

$$f_p = 9\sqrt{n(10^{12}\,\text{cm}^{-3})}\ \text{GHz} \qquad (9)$$

This is called the *plasma frequency*, and it depends only on the plasma density.

The sound wave in the ion fluid behaves quite differently. It has a characteristic velocity rather than a characteristic frequency, and the frequency, of course, is much lower. The physical difference is that, as the ions are displaced from their equilibrium positions, the more mobile electrons can move with them to shield out their charges. However, the shielding is not perfect because the electron have thermal motions which are random and allow a small electric field to leak out of the Debye cloud. These *ion acoustic waves*, or simply *ion waves*, propagate with the *ion acoustic velocity* or *ion sound speed c_s*:

$$c_s \equiv \left(KT_e / M \right)^{1/2} \qquad (10)$$

where capital M is the ion mass. Note that c_s depends on T_e, not T_i, as in air, because the deviation from perfect Debye shielding depends on T_e. There is actually a small correction $\propto T_i$ which we have neglected because T_i is normally $<< T_e$ in partially ionized plasmas. The hybrid ratio T_e / M permits ion sound waves to exist even when the ions are cold.

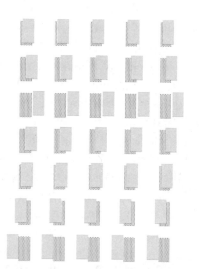

A plasma oscillation: displaced electrons oscillate around fixed ions. The wave does not necessarily propagate.

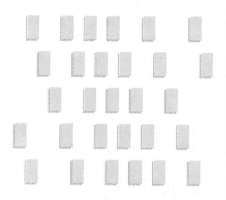

An ion acoustic wave: ions and electrons move together in a propagating compressional wave.

3. Larmor radius and cyclotron frequency

If the plasma is imbedded in a DC magnetic field (B-field), many more types of wave motions are possible than those given in the previous section. This is because the B-field affects the motions of the charged particles and makes the plasma an anisotropic medium, with a preferred direction along **B**. As long as the ion or electron of charge q is moving, it feels a Lorentz force $q\mathbf{v} \times \mathbf{B}$, which is perpendicular to the both the velocity and the field. This force has no effect on the velocity component parallel to **B**, but in the perpendicular plane it forces the particle to gyrate in a *cyclotron orbit*. The frequency of this circular motion, the *cyclotron frequency* ω_c, is independent of velocity and depends only on the charge-to-mass ratio:

$$\boxed{\omega_c = |qB|/m}, \quad \text{or} \quad \boxed{f_c = \omega_c/2\pi \approx 2.8 \text{ MHz/G}} \quad (11)$$

The radius of the circle of gyration, called *Larmor radius* or *gyroradius* r_L, however, does depend on velocity. If $v_\perp$ is the velocity component in the plane perpendicular to **B**, a particle completes an orbit of length $2\pi r_L$ in a time $2\pi/\omega_c$, so $v_\perp = r_L\omega_c$, or

$$\boxed{r_L = v_\perp / \omega_c} \quad (12)$$

Since $\omega_c \propto 1/M$ while $v_\perp \propto 1/M^{1/2}$, r_L tends to be smaller for electrons than for ions by the square root of the mass ratio. In processing plasmas that have magnetic fields, the fields are usually of the order of several hundred gauss ($1G = 10^{-4}$ Tesla), in which case heavy ions such as Cl are not much affected by **B**, while electrons are strongly constrained to move along **B**, while gyrating rapidly in small circles in the perpendicular plane. In this case, if is often possible to neglect the small gyroradius and treat only the motion of the center of the orbit, called the *guiding center*. Note that ions and electrons gyrate in opposite directions. An easy way to remember the direction is to consider the moving charge as a current, taking into account the sign of the charge. This current generates a magnetic field in a direction given by the right-hand rule, and the current must always be in a direction so as to generate a magnetic field opposing the background magnetic field.

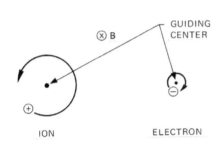

Electrons and ions gyrate in opposite directions with different size orbits.

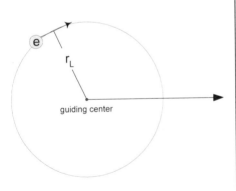

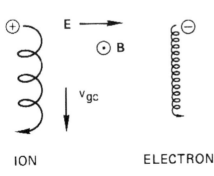

The **E** × **B** drift

4. E × B drift

In magnetic fields so strong that both ions and electrons have Larmor radii much smaller than the plasma radius, the particles' guiding centers drift across **B** in response to applied electric fields $\mathbf{E}_\perp$ (the component perpendicular to **B**). This drift speed is given by

$$v_E = \mathbf{E} \times \mathbf{B} / B^2. \qquad (13)$$

The velocity parallel to **B** is, of course, unaffected by $\mathbf{E}_\perp$. Note that v_E is perpendicular to both **E** and **B** and is the same for ions and electrons. If **E** is not constant across an ion Larmor diameter, the ions feel an average E-field and tend to drift somewhat more slowly than the electrons. At fields of a few hundred gauss, as is common in plasma processing, heavy ions such as argon or chlorine may strike the wall before completing a Larmor orbit, especially if they have been accelerated to an energy $>> KT_i$ by $\mathbf{E}_\perp$. In that case, one has a hybrid situation in which the ions are basically unmagnetized, while the electrons are strongly magnetized and follow Eq. (13).

5. Sheaths and presheaths

We come now to the details of how a sheath is formed. Let there be a wall at $x = 0$, with a plasma extending a large distance to the right ($x > 0$). At $x = s$ we draw an imaginary plane which we can call the *sheath edge*. From our discussion of Debye shielding, we would expect s to be of the order of λ_D (actually, it is more like $5\lambda_D$).. Outside the sheath ($x \geq s$), quasineutrality requires $n_i \approx n_e$. Let the plasma potential there be defined as $V = 0$. Inside the sheath, we can have an imbalance of charges. The potential in the sheath must be negative in order to repel electrons, and this means that $V(x)$ must have negative curvature.. From the one-dimensional form of Eq. (3), we see that n_i must be larger than n_e. Now, if the electrons are Maxwellian, their density in a potential hill will be exponentially smaller:

$$n_e / n_s = \exp(eV/KT_e), \qquad (14)$$

where n_s is the density at the sheath edge. To calculate the ion density, consider that the ions flowing toward the wall are accelerated by the sheath's E-field and are not reflected, so the ion flux is constant. We may neglect T_i, but for reasons that will become clear, we have to assume that the ions enter the sheath with a finite velocity v_s. The equation of continuity is then

$$n_i v_i = n_s v_s \qquad (15)$$

Conservation of energy gives

$$\tfrac{1}{2} M v_i^2 + eV = \tfrac{1}{2} M v_s^2 \qquad (16)$$

The last two equations give

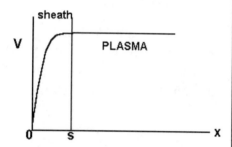

The sheath potential can have the proper curvature only if $n_i > n_e$ there.

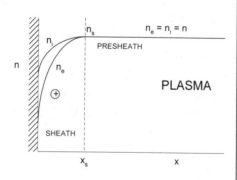

Only in the sheath can quasineutrality be violated.

$$\frac{n_i}{n_s} = \frac{1}{\left[1 - 2eV / Mv_s^2\right]^{1/2}} \tag{17}$$

The sheath condition $n_i > n_e$ has to hold even for small values of $|V|$, just inside the sheath. In that case, we can expand Eqs. (14) and (17) in Taylor series to obtain

$$\frac{n_e}{n_s} = 1 + \frac{eV}{KT_e} + \ldots, \qquad \frac{n_i}{n_s} = 1 + \frac{eV}{Mv_s^2} + \ldots \tag{18}$$

Since V is negative, the condition $n_i > n_e$ then becomes

$$\boxed{\frac{e|V|}{Mv_s^2} < \frac{e|V|}{KT_e}} \tag{19}$$

The sheath condition is then

$$\boxed{v_s > (KT_e / M)^{1/2} = c_s} \tag{20}$$

This is called the *Bohm sheath criterion* and states that ions must stream into the sheath with a velocity at least as large as the acoustic velocity in order for a sheath of the right shape to form. Such a *Debye sheath* is also called an *ion sheath*, since it has a net positive charge.

The obvious question now is: "How can the ions get such a large directed velocity, which is much larger than their thermal energies?" There must be a small electric field in the quasineutral region of the main body of the plasma that accelerates ions to an energy of at least $\frac{1}{2}KT_e$ toward the sheath edge. Such an E-field can exist only by virtue of non-ideal effects: collisions, ionization, or other sources of particles or momentum. This region is called the *presheath*, and it extends over distances of the order of the plasma dimensions. The pre-sheath field is weak enough that quasineutrality does not have to be violated to create it. In reference to plasma processing, we see that ions naturally gain a directed velocity by the time they strike the substrate, even if nothing is done to enhance the sheath drop. If a voltage is applied between two walls or electrodes, there will still be an ion sheath on each wall, but the sheath drops will be unequal, so the electron fluxes to each wall will be unequal even if they have the same area. However, the ion fluxes are the same ($= n_s v_s$) to each wall, and the total electron flux must equal the total ion flux. Since more electrons are collected at the more positive electrode than at the other, a current has to flow through the biasing power supply.

If the sheaths drops are unequal, the electron fluxes will be unequal, but they must add up to the total ion flux (which is the same to both sides).

If a presheath has to exist, the density n_s at the sheath edge cannot be the same as the plasma density n in the body of the plasma. Since the ions have a velocity c_s at the sheath edge, their energy $\frac{1}{2}Mc_s^2$ is $\frac{1}{2}KT_e$, and there must be a potential drop of at least $\frac{1}{2}KT_e$ between the body of the plasma and the sheath edge. Let us now set $V = 0$ inside the main plasma, so that $V = V_s$ at the sheath edge. The electrons are still assumed to be in a Maxwellian distribution:

$$n_e = n_0 e^{eV / KT_e} . \qquad (21)$$

Since the integral of an exponential is still an exponential, it is the property of a Maxwellian distribution that it remains a Maxwellian at the same temperature when placed in a retarding potential; only the density is changed. There is only a small modification in the number of electrons moving back from the sheath due to the few electrons that are lost through the Coulomb barrier. Thus, Eq. (21) holds throughout the plasma, presheath, and sheath, regardless of whether there are collisions or not. If $eV_s = -e|V_s| = -\frac{1}{2}KT_e$, then Eq. (21) tells us that

$$n_s = n_0 e^{-1/2} = 0.6n_0 \approx \frac{1}{2}n_0 . \qquad (22)$$

This is approximate, since there is no sharp dividing line between sheath and presheath. In the future we shall use the simple relation $n_s \approx \frac{1}{2}n_0$, where n_0 is the density in the main plasma.

In summary, a plasma can coexist with a material boundary only if a thin sheath forms, isolating the plasma from the boundary. In the sheath there is a *Coulomb barrier*, or potential drop, of magnitude several times KT_e, which repels electrons from and accelerates ions toward the wall. The sheath drop adjusts itself so that the fluxes of ions and electrons leaving the plasma are almost exactly equal, so that quasineutrality is maintained.

PRINCIPLES OF PLASMA PROCESSING
Course Notes: Prof. F.F. Chen

PART A2: INTRODUCTION TO GAS DISCHARGES

III. GAS DISCHARGE FUNDAMENTALS

1. Collision cross sections and mean free path (Chen, p.155*ff*)[*]

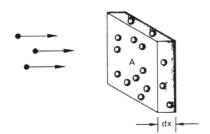

Definition of cross section

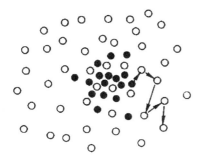

Diffusion is a random walk process.

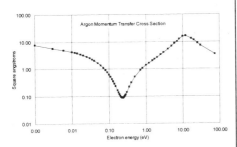

Momentum transfer cross section for argon, showing the Ramsauer minimum

We consider first the collisions of ions and electrons with the neutral atoms in a partially ionized plasma; collisions between charged particles are more complicated and will be treated later. Since neutral atoms have no external electric field, ions and electrons do not feel the presence of a neutral until they come within an atomic radius of it. When an electron, say, collides with a neutral, it will bounce off it most of the time as if it were a billiard ball. We can then assign to the atom an effective cross sectional area, or *momentum transfer cross section*, which means that, on the average, an electron hitting such an area around the center of an atom would have its (vector) momentum changed by a lot; a lot being a change comparable to the size of the original momentum. The cross section that an electron sees depends on its energy, so in general a cross section σ depends on the energy, or, on average, the temperature of the bombarding particles. Atoms are about 10^{-8} cm (1 Angstrom) in radius, so atomic cross sections tend to be around 10^{-16} cm^2 (1 Å^2) in magnitude. People often express cross sections in units of $\pi a_0^2 = 0.88 \times 10^{-16}$ cm^2, where a_0 is the radius of the hydrogen atom.

At high energies, cross sections tend to decrease with energy, varying as $1/v$, where v is the velocity of the incoming particle. This is because the electron goes past the atom so fast that there is not enough time for the electric field of the outermost electrons of the atom to change the momentum of the passing particle. At low energies, however, $\sigma(v)$ can be more constant, or can even go up with energy, depending on the details of how the atomic fields are shaped. A famous case is the Ramsauer cross section, occurring for noble gases like argon, which takes a deep dive around 1 eV. Electrons of such low energies can almost pass through a Ramsauer atom without knowing it is there.

Ions have somewhat higher cross sections with

---

[*] References are for further information if you need it.

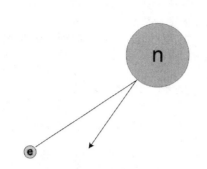

An elastic electron-neutral collision

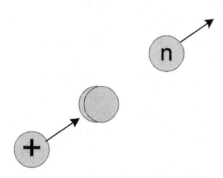

An ion-neutral charge exchange
collision

neutrals because the similarity in mass makes it easier for the ion to exchange momentum with the neutral. Ions colliding with neutrals of the same species, such as Cl with Cl^+, have a special effect, called a *charge exchange collision*. A ion passing close to an atom can pull off an outer electron from the atom, thus ionizing it. The ion then becomes a fast neutral, while the neutral becomes a slow ion. There is no large momentum exchange, but the change in identity makes it look like a huge collision in which the ion has lost most of its energy. Charge-exchange cross sections (σ_{cx}) can be as large as $100\ \pi a_0^2$.

Unless one is dealing with a monoenergetic beam of electrons or ions, a much more useful quantity is the collision probability $<\sigma v>$, measured in cm^3/sec, where the average is taken over a Maxwellian distribution at temperature KT_e or KT_i. The average rate at which each electron in that distribution makes a collision with an atom is then $<\sigma v>$ times the density of neutrals; thus, the collision frequency is:

$$v_c = n_n <\sigma v > \text{ per sec.} \qquad (1)$$

If the density of electrons is n_e, the number of collisions per cm^3/sec is just

$$n_e n_n <\sigma v > \ cm^{-3} \ sec^{-1}. \qquad (2)$$

The same rate holds for ion-neutral collisions if the appropriate ion value of $<\sigma v>$ is used. On average, a particle makes a collision after traveling a distance λ_m, called the *mean free path*. Since distance is velocity times time, dividing v by Eq. (1) (before averaging) gives

$$\lambda_m = 1/n_n \sigma. \qquad (3)$$

This is actually the mean free path for each velocity of particle, not the average mean free path for a Maxwellian distribution.

2. Ionization and excitation cross sections (L & L, Chap. 3).

If the incoming particle has enough energy, it can do more than bounce off an atom; it can disturb the electrons orbiting the atom, making an *inelastic collision*. Sometimes only the outermost electron is kicked into a higher energy level, leaving the atom in an excited state. The atom then decays spontaneously into a *metastable* state or back to the ground level, emitting a photon of a particular energy or wavelength. There is an *excitation cross section* for each such transition or each spectral line that is characteristic of that atom. Electrons of higher energy can knock an electron off the atom entirely, thus *ionizing* it. As every freshman physics student knows, it takes 13.6 eV to ionize a hydrogen atom; most other atoms have ionization thresholds slightly higher than this value. The frequency of ionization is related by Eq. (3) to the *ionization cross section* σ_{ion}, which obviously is zero below the threshold energy E_{ion}. It increases rapidly above E_{ion}, then tapers off around 50 or 100 eV and then decays at very high energies because the electrons zip by so fast that their force on the bound electrons is felt only for a very short time. Since only a small number of electrons in the tail of a 4-eV distribution, say, have enough energy to ionize, σ_{ion} increases exponentially with KT_e up to temperatures of 100 eV or so.

Double ionizations are extremely rare in a single collision, but a singly ionized atom can be ionized in another collision with an electron to become doubly ionized; for instance $Ar^+ \rightarrow Ar^{++}$. Industrial plasmas are usually cool enough that almost all ions are only singly charged. Some ions have an affinity for electrons and can hold on to an extra one, becoming a negative ion. Cl^- and the molecule SF_6^- are common examples. There are electron *attachment cross sections* for this process, which occurs at very low electron temperatures.

3. Coulomb collisions; resistivity (Chen, p. 176*ff*).

Now we consider collisions between charged particles (*Coulomb collisions*). We can give a physical description of the action and then the formulas that will be useful, but the derivation of these formulas is beyond our scope. When an electron collides with an ion, it feels the electric field of the positive ion from a distance and is gradually pulled toward it. Conversely, an electron can feel the repelling field of another electron when it is many atomic radii away. These particles are basically point charges, so they do not actually collide; they swing around one another and change their trajectories. We can define an effective cross section as πh^2, where h is the *impact parameter* (the distance the particle would miss its target by if it went straight) for which the trajectory is deflected by 90°. However, this is not the real cross section, because there is Debye shielding. A cloud of negative charge is attracted around any positive charge and shields out the electric field so that it is much weaker at large distances than it would otherwise be. This Debye cloud has a thickness of order λ_D. The amount of potential that can leak out of the Debye cloud is about $\frac{1}{2}KT_e$ (see the discussion of presheath in Sec. II-5). Because of

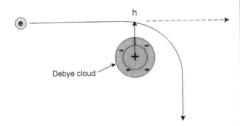

A 90° electron-ion collision

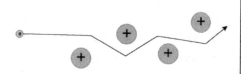

Electrons "collide" via numerous
small-angle deflections.

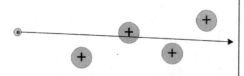

Fast electrons hardly collide at all.

tial that can leak out of the Debye cloud is about $\frac{1}{2}KT_e$ (see the discussion of presheath in Sec. II-5). Because of this shielding, incident particles suffer only a small change in trajectory most of the time. However, there are many such small-angle collisions, and their cumulative effect is to make the effective cross section larger. This effect is difficult to calculate exactly, but fortunately the details make little difference. The 90° cross section is to be multiplied by a factor $\ln \Lambda$, where Λ is the ratio λ_D/h. Since only the logarithm of Λ enters, one does not have to evaluate Λ exactly; **ln Λ can be approximated by 10** in almost all situations we shall encounter. The resulting approximate formulas for the electron-ion and electron-electron collision frequencies are, respectively,

$$\nu_{ei} \approx 2.9 \times 10^{-6} n_{cm} \ln \Lambda / T_{eV}^{3/2}$$
$$\nu_{ee} \approx 5.8 \times 10^{-6} n_{cm} \ln \Lambda / T_{eV}^{3/2} \, , \qquad (4)$$

where n_{cm} is in cm^{-3}, T_{eV} is KT_e in eV, and $\ln\Lambda \approx 10$. There are, of course, many other types of collisions, but these formulas are all we need most of the time.

Note that these frequencies depend only on T_e, because the ions' slight motion during the collision can be neglected. The factor n on the right is of course the density of the targets, but for singly charged ions the ion and electron densities are the same. Note also that the collision frequency varies as $KT_e^{-3/2}$, or on v^{-3}. For charged particles, the collision rate decreases much faster with temperature than for neutral collisions. In hot plasmas, the particles collide so infrequently that we can consider the plasma to be collisionless.

The resistivity of a piece of copper wire depends on how frequently the conduction electrons collide with the copper ions as they try to move through them to carry the current. Similarly, plasma has a resistivity related to the collision rate ν_{ei} above. The specific resistivity of a plasma is given by

$$\eta = m\nu_{ei} / ne^2 \, . \qquad (5)$$

Note that the factor n cancels out because $\nu_{ei} \propto n$. **The plasma resistivity is independent of density.** This is because the number of charge carriers increases with density, but so does the number of ions which slow them down. In practical units, resistivity is given by

$$\eta_{\parallel} = 5.2 \times 10^{-5} Z \ln \Lambda / T_{eV}^{3/2} \quad \Omega - m. \qquad (6)$$

Here we have generalized to ions of charge Z and have added a parallel sign to η in anticipation of the magnetic

4. Transition between neutral- and ion-dominated electron collisions

The behavior of a partially ionized plasma depends a great deal on the collisionality of the electrons. From the discussion above, we can compute their collision rate against neutrals and ions. Collisions between electrons themselves are not important here; these just redistribute the energies of the electrons so that they remain in a Maxwellian distribution.

The collision rate between electrons and neutrals is given by

$$v_{en} = n_n < \sigma v >_{en}, \tag{7}$$

where the σ is the total cross section for *e-n* collisions but can be approximated by the elastic cross section, since the inelastic processes generally have smaller cross sections. The neutral density n_n is related to the fill pressure n_{n0} of the gas. It is convenient to measure pressure in Torr or mTorr. A Torr of pressure supports the weight of a 1-mm high column of Hg, and atmospheric pressure is 760 Torr. A millitorr (mTorr) is also called a *micron* of pressure. Some people like to measure pressure in Pascals, where 1 Pa = 7.510 mTorr, or about 7 times as large as a mTorr. At 20°C and pressure of p mTorr, the neutral density is

$$\boxed{n_n \approx 3.3 \times 10^{13} \, p(\text{mTorr}) \quad \text{cm}^{-3}}. \tag{8}$$

If this were all ionized, the plasma density would be $n_e = n_i = n = n_{n0}$, but only for a monatomic gas like argon. A diatomic gas like Cl_2 would have $n = 2n_{n0}$. Are e-i collisions as important as e-n collisions? To get a rough estimate of v_{en}, we can take $<\sigma v>$ to be $<\sigma><v>$, σ to be $\approx 10^{-16}$ cm^2, and $<v>$ to be the thermal velocity v_{th}, defined by

$$v_{th} \equiv (2KT/m)^{1/2},$$
$$v_{th,e} = (2KT_e/m)^{1/2} \approx 6 \times 10^7 T_{eV}^{1/2} \text{ cm/sec} \tag{9}$$

We then have

$$v_{en} \approx (3.3 \times 10^{13}) p \bullet (10^{-16}) \bullet 6 \times 10^7 T_{eV}^{1/2}$$
$$\approx 2 \times 10^5 p_{mTorr} T_{eV}^{1/2} \tag{10}$$

(This formula is an order-of-magnitude estimate and is not to be used in exact calculations.) The electron-ion collision frequency is given by Eq. (4):

$$\nu_{ei} \approx 2.9 \times 10^{-5} n / T_{eV}^{3/2} \ . \tag{11}$$

The ratio then gives

$$\frac{\nu_{ei}}{\nu_{en}} \approx 1.5 \times 10^{-10} \frac{n}{p} T_{eV}^{-2} \ . \tag{12}$$

The crossover point, when this ratio is unity, occurs for a density of

$$n_{crit} \approx 6.9 \times 10^{9} \, p_{mTorr} T_{eV}^{2} \ cm^{-3} \ . \tag{13}$$

For instance, if $p = 3$ mTorr and $KT_e = 3$ eV, the crossover density is $n_{crit} = 1.9 \times 10^{11}$ cm^{-3}. Thus, High Density Plasma (HDP) sources operating in the high 10^{11} to mid-10^{12} cm^{-3} range are controlled by electron-ion collisions, while older low-density sources such as the RIE operating in the 10^{10} to mid-10^{11} cm^{-3} range are controlled by electron-neutral collisions. The worst case is in between, when both types of collisions have to be taken into account.

5. Mobility, diffusion, ambipolar diffusion (Chen, p.155*ff*)

Now that we know the collision rates, we can see how they affect the motions of the plasma particles. If we apply an electric field **E** (V/m) to a plasma, electrons will move in the −**E** direction and carry a current. For a fully ionized plasma, we have seen how to compute the specific resistivity η. The current density is then given by

$$\mathbf{j} = \mathbf{E} / \eta \quad A / m^2 \tag{14}$$

In a weakly ionized gas, the electrons will come to a steady velocity as they lose energy in neutral collisions but regain it from the E-field between collisions. This average drift velocity is of course proportional to E, and the constant of proportionality is called the *mobility* μ, which is related to the collision frequency:

$$\mathbf{u} = -\mu\mathbf{E}, \quad \mu_e = e / m\nu_{en} \ . \tag{15}$$

By e we always mean the *magnitude* of the elementary charge. There is an analogous expression for ion mobility, but the ions will not carry much current. The *flux* of electrons Γ_e and the corresponding current density are given by

$$\Gamma_e = -n_e\mu_e\mathbf{E}, \quad \mathbf{j} = en_e\mu_e\mathbf{E}, \tag{16}$$

and similarly for ions. How do these E-fields get into the

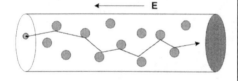

Conductivity is determined by the average drift velocity **u** that an electron gets while colliding with neutrals or ions. In a wire, the number of target atoms is unrelated to the number of charge carriers, but in a plasma, the ion and electron densities are equal.

plasma when there is Debye shielding? If one applies a voltage to part of the wall or to an electrode inside the plasma, electrons will move so as to shield it out, but because of the presheath effect a small electric field will always leak out into the plasma. The presheath field can be large only at high pressures. To apply larger E-fields, one can use *inductive coupling*, in which a time-varying magnetic field is imposed on the plasma by external antennas or coils, and this field induces an electric field by Faraday's Law. Electron *currents* in the plasma will still try to shield out this induced field, but in a different way; magnetic fields can reduce this shielding. We shall discuss this further under Plasma Sources.

The plasma density will usually be nonuniform, being high in the middle and tapering off toward the walls. Each species will diffuse toward the wall; more specifically, toward regions of lower density. The diffusion velocity is proportional to the density gradient ∇n, and the constant of proportionality is the diffusion coefficient D:

$$\mathbf{u} = -D\nabla n / n, \quad D_e = KT_e / mv_{en}, \qquad (17)$$

and similarly for the ions. The diffusion flux is then given by

$$\Gamma = -D\nabla n. \qquad (18)$$

Note that D has dimensions of an area, and Γ is in units of number per square meter per second.

The sum of the fluxes toward the wall from mobility and diffusion is then

$$\Gamma_e = -n\mu_e\mathbf{E} - D_e\nabla n$$
$$\Gamma_i = +n\mu_i\mathbf{E} - D_i\nabla n \qquad (19)$$

Note that the sign is different in the mobility term. Since μ and D are larger for electrons than for ions, Γ_e will be larger than Γ_i, and there will soon be a large charge imbalance. To stay quasi-neutral, an electric field will naturally arise so as to speed up the diffusion of ions and retard the diffusion of electrons. This field, called the *ambipolar* field, exists in the body of the plasma where the collisions occur, not in the sheath. To calculate this field, we set $\Gamma_e = \Gamma_i$ and solve for $\mathbf{E}$. Adding and subtracting the equations in (19), we get

$$\Gamma_i + \Gamma_e \equiv 2\Gamma_a = n(\mu_i - \mu_e)\mathbf{E} - (D_i + D_e)\nabla n$$
$$\Gamma_i - \Gamma_e \equiv 0 = n(\mu_i + \mu_e)\mathbf{E} - (D_i - D_e)\nabla n \qquad (20)$$

From these we can solve for the ambipolar flux Γ_a, ob-

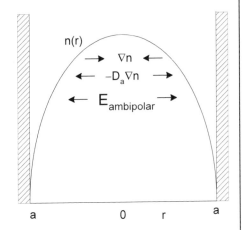

taining

$$\Gamma_a = -\frac{\mu_i D_e + \mu_e D_i}{\mu_i + \mu_e} \nabla n \equiv -D_a \nabla n . \qquad (21)$$

We see that diffusion with the self-generated E-field, called *ambipolar diffusion*, follows the usual diffusion law, Eq. (18), but with an *ambipolar diffusion coefficient* D_a defined in Eq. (21). Since, from (15) and (17), μ and D are related by

$$\mu = eD / KT , \qquad (22)$$

and μ_e is usually much greater than μ_i, D_a is well approximated by

$$D_a \approx \left(\frac{T_e}{T_i} + 1\right) D_i \approx \frac{T_e}{T_i} D_i , \qquad (23)$$

meaning that the loss of plasma to the walls is slowed down to the loss rate of the slower species, modified by the temperature ratio.

6. Magnetic field effects; magnetic buckets (Chen, p. 176*ff*)

Diffusion of plasma in a magnetic field is complicated, because particle motion is anisotropic. If there were no collisions and the cyclotron orbits were all smaller than the dimensions of the container, ions and electrons would not diffuse across **B** at all. They would just spin in their Larmor orbits and move freely in the **z** direction (the direction of **B**). But when they collide with one another or with a neutral, their guiding centers can get shifted, and then there can be cross-field diffusion. First, let us consider charged-neutral collisions. The transport coefficients $D_\parallel$ and $\mu_\parallel$ *along* **B** are unchanged from Eqs. (15) and (17), but the coefficients *across* **B** are changed to the following:

$$D_\perp = \frac{D_\parallel}{1 + (\omega_c / \nu_c)^2} , \qquad \mu_\perp = \frac{\mu_\parallel}{1 + (\omega_c / \nu_c)^2} ,$$

$$D_\parallel = \frac{KT}{m\nu_c} , \qquad \mu_\parallel = \frac{e}{m\nu_c} \qquad (24)$$

Here ν_c is the collision frequency against neutrals, and we have repeated the parallel definitions for convenience. It is understood that all these parameters depend on species, ions or electrons. If the ratio ω_c/ν_c is small, the magnetic field has little effect. When it is large, the particles are *strongly magnetized*. When ω_c/ν_c is of order unity, we have the in-between case. If σ and KT are the

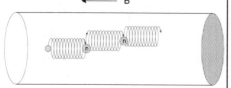

Diffusion of an electron across a magnetic field

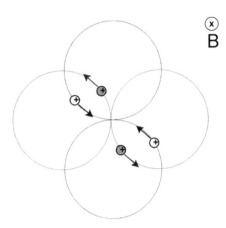

Ⓧ B

Like-particles collisions do not cause diffusion, because the orbits after the collision (dashed lines) have guiding centers that are simply rotated.

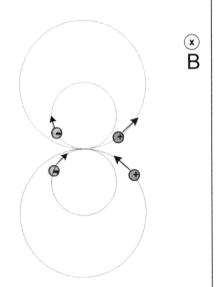

Ⓧ B

Collisions between positive and negative particles cause both guiding centers to move in the same direction, resulting in cross-field diffusion.

same, electrons have ω_c/v_c values $\sqrt{(M/m)}$ times larger, and their Larmor radii are $\sqrt{(M/m)}$ times smaller than for ions (a factor of 271 for argon). So in B-fields of 100-1000 G, as one might have in processing machines, electrons would be strongly magnetized, and ions perhaps weakly magnetized or not magnetized at all. If ω_c/v_c is large, the "1" in Eq. (24) can be neglected, and we see that $D_\perp \propto v_c$, while $D_\parallel \propto 1/v_c$. Thus, collisions impede diffusion along **B** but increases diffusion across **B**.

We now consider collisions between strongly magnetized charged particles. It turns out that like-like collisions—that is, ion-ion or electron-electron collisions—do not produce any appreciable diffusion. That is because the two colliding particles have a center of mass, and all that happens in a collision is that the particles shift around relative to the center of mass. The center of mass itself doesn't go anywhere. This is the reason we did not need to give the ion-ion collision frequency v_{ii}. However, when an electron and an ion collide with each other, *both their gyration centers move in the same direction*. The reason for this can be traced back to the fact that the two particles gyrate in opposite directions. So collisions between electrons and ions allow cross-field diffusion to occur. However, the cross-field mobility is zero, in the lowest approximation, because the v_E drifts are equal. Consider what would happen if an ambipolar field were to build up in the radial direction in a cylindrical plasma. An E-field across **B** cannot move guiding centers along **E**, but only in the **E** × **B** direction (Sec. II-4). If ions and electrons were to diffuse at different rates toward the wall, the resulting space charge would build up a radial electric field of such a sign as to retard the faster-diffusing species. But this E-field cannot slow up those particles; it can only spin them in the azimuthal direction. Then the plasma would spin faster and faster until it blows up. Fortunately, this does not happen because the ion and electron diffusion rates *are the same* across **B** in a fully ionized plasma. This is not a coincidence; it results from momentum conservation, there being no third species (neutrals) to take up the momentum. In summary, for a fully ionized plasma there is no cross-field mobility, and the cross-field diffusion coefficient, the same for ions and electrons, is given by:

$$D_{\perp c} = \frac{\eta_\perp n(KT_i + KT_e)}{B^2} \quad . \tag{25}$$

Here $\eta_\perp$ is the transverse resistivity, which is about twice as large at that given in Eq. (5).

Note that we have given the label "c" to $D_\perp$, standing for "classical". This is because electrons do not always behave the way classical theory would predict; in fact, they almost never do. Electrons are so mobile that they can find other ways to get across the magnetic field. For instance, they can generate bursts of plasma oscillations, of such high frequency that one would not notice them, to move themselves by means of the electric fields of the waves. Or they can go along the B-field to the end of the discharge and then adjust the sheath drop there so as to change the potential along that field line and change the transverse electric fields in the plasma. This is one of the problems in controlled fusion; it has not yet been solved. Fortunately, ions are so slow that they have no such anomalous behavior, and they can be depended upon to move classically.

In processing plasmas that have a magnetic field, electrons are strongly magnetized, but ions are almost unmagnetized. What do we do then? For parallel diffusion, the formulas are not affected. For transverse diffusion, we can use $D_{\perp e}$ for electrons and $D_{\|i}$ for ions, but there in no rigorous theory for this. Plasma processing is so new that problems like this are still being researched.

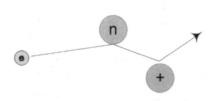

If the ions are weakly magnetized, electrons-ion collisions can be treated like electron-neutral collisions, but with a different collision frequency.

Finally, we come to "magnetic buckets," which were invented at UCLA and are used in some plasma reactors. A magnetic bucket is a chamber in which the walls are covered with a localized magnetic field existing only near the surface. This field can be made with permanent magnets held in an array outside the chamber, and it has the shape of a "picket fence", or multiple cusps (Chen, cover illustration). The idea is that the plasma is free to diffuse and make itself uniform inside the bucket, but when it tries to get out, it is impeded by the surface field. However, the surface field has leaks in it, and cool electrons are collisional enough to get through these leaks. One would not expect the fence to be very effective against loss of the bulk electrons. However, the "primary" electrons, the ones that have enough energy to ionize, are less collisional and may be confined in the bucket. There has been no definitive experiment on this, but in some reactors magnetic buckets have been found to confine plasmas better as they stream from the source toward the wafer.

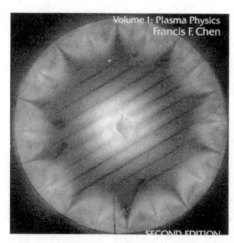

Light emission excited by fast electrons shows the shape of the field lines in a magnetic bucket.

The following graphs provide cross section data for the homework problems.

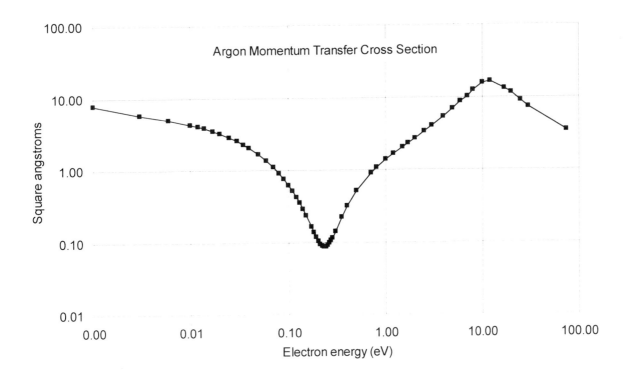

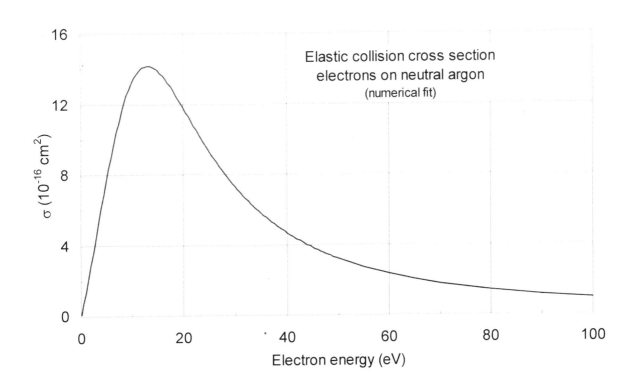

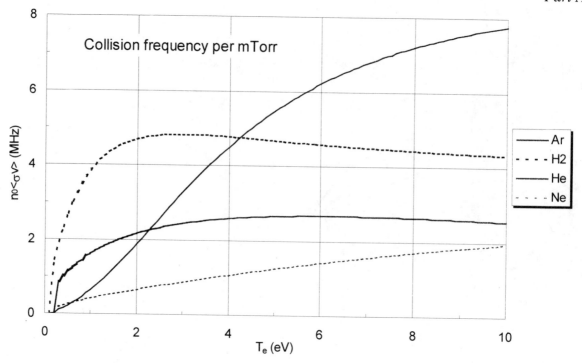

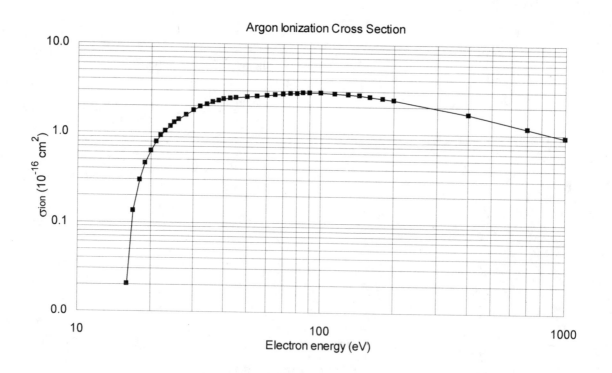

Ionization probability in argon

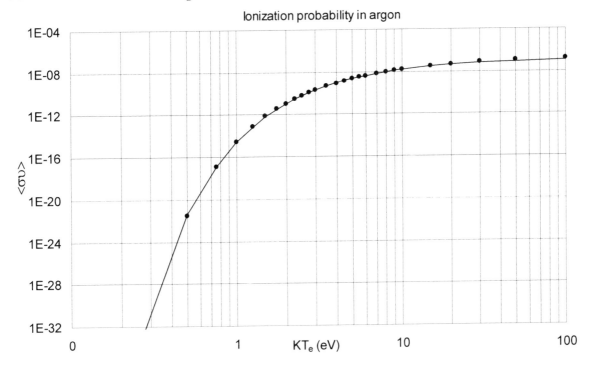

Ionization cross sections

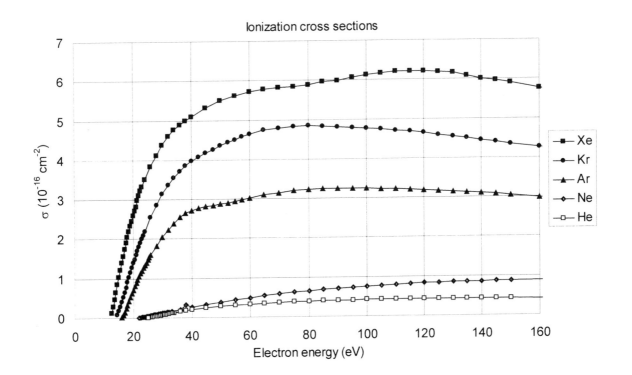

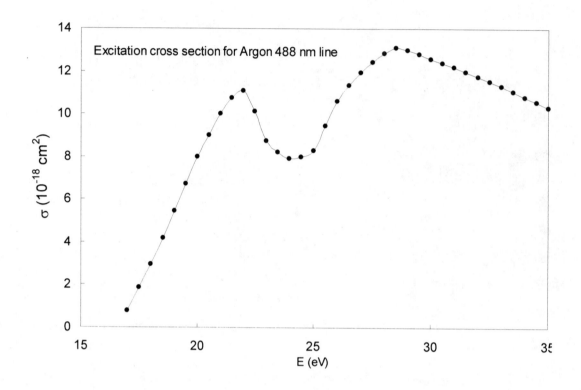

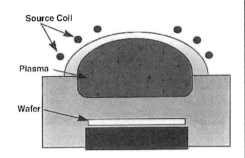

A typical plasma reactor

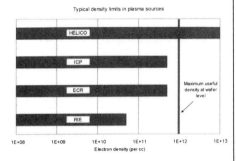

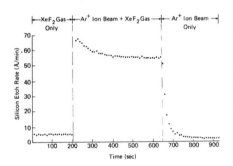

Coburn's famous graph shows that the etch rate is greatly enhanced when a plasma is added. On left: only chemical etching. On right: only plasma sputtering.

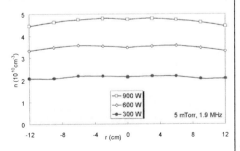

Typical uniformity in a reactor.

PRINCIPLES OF PLASMA PROCESSING
Course Notes: Prof. F.F. Chen

PART A3: PLASMA SOURCES I

IV. INTRODUCTION TO PLASMA SOURCES

1. Desirable characteristics of plasma processing sources

The ideal plasma generator would excel in all of the following characteristics, but some compromises are always necessary. Advanced plasma tools are in production that satisfy these criteria quite well. What is important, however, is the combination of the tool and the *process*. For instance, etching SiO_2 requires both a source and a procedure. The commercial product is often not just the tool but the process, including the source, the settings, and the timing developed to perform a given task.

- **Etch rate.** High etch rate normally requires high plasma density. Some experiments have shown that, more exactly, the etch rate is proportional to the ion energy flux; that is, to the ion flux to the wafer times the average energy of the ions. High etch rate is especially important in the fabrication of MEMS (MicroElectroMechanical Systems), where large amounts of material has to be removed.

- **Uniformity.** To process a wafer evenly from center to edge requires a plasma that is uniform in density, temperature, and potential. Computer chips near the edge of a wafer often suffer from substandard processing, resulting in a lower speed rating for those CPUs.

- **Anisotropy.** To etch straight trench walls, the ions must impinge on the wafer at normal incidence; this is called *anisotropy*. To achieve this, the sheath edge must be planar all the way across the wafer.

- **Selectivity.** By this we mean the ability to etch one material faster than another. Polysilicon etches faster than SiO_2. To etch SiO_2 preferentially requires a fortuitous series of events. There is always deposition of hydrocarbon polymers during the etching process, and these inhibit further etching. Both poly and oxide are covered, but the polymer layer is more easily removed from SiO_2 because of the oxygen that is released from SiO_2. The polymer layer prevents further etching of the silicon. A critical problem is the photoresist/polysilicon selectivity, which currently has a low value around 5. Increasing

Deep Trench Etch
256MB DRAM Device

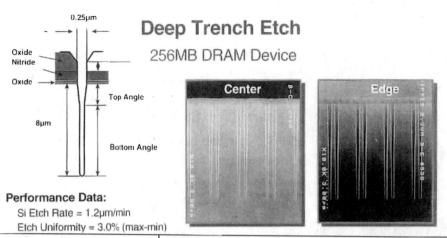

Performance Data:
Si Etch Rate = 1.2μm/min
Etch Uniformity = 3.0% (max-min)

Anisotropy permits etching downwards without going sideways. These deep trenches actually require help from polymers deposited on the sides. Source: Applied Materials.

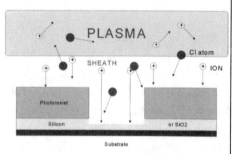

Selectivity allows overetching without cutting into the next layer.

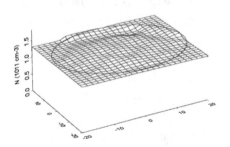

A 2D plot of plasma density shows uniform coverage of a large area.

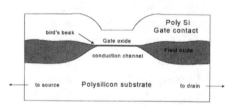

Plasma etching tends to build up large voltages across thin insulators, damaging them. This is a serious problem

this number would alleviate deformation of the mask during processing. Because of these indirect effects, it is not clear what properties of the plasma source control selectivity. One hopes that by altering the electron velocity distribution $f(v)$, one could change the chemical precursors in such a way as to control selectivity.

- **Area coverage.** The semiconductor industry started with Si wafers of 4-inch diameter, gradually increasing to 6, 8, and 12 inches. Current production is based on 200-mm (8-inch) wafers, and the plan is to retool both ingot factories and fabs for 300-mm wafers. More chips can be produced at once with these large wafers, since the size of each chip—the *die size*—is kept relatively constant. Plasma sources have already been developed to cover 12-in wafers uniformly. The flat-panel display industry, however, uses glass substrates as large as 600 by 900 mm. Plasma sources of this size are now used for deposition, but low-pressure sources for etching will be needed in the future.

- **Low damage.** Thin oxide layers are easily damaged during plasma processing, and this is a serious problem for the industry. Nonuniform sheath drops and magnetic fields near the wafer have been shown to increase damage, but these problems are under control with current plasma tools. Damage by energetic ion bombardment and UV radiation are lesser effects compared with *electron shading*. The latter occurs when ions but not electrons reach the bottom of a trench being etched, causing a charge buildup which drives current through the insulating oxide layer. There has been considerable evidence that low T_e will minimize electron shading damage, but the picture is far from clear.

- **Adaptability.** Since each process requires a different gas mixture, pressure, power level, etc., plasma sources should be able to operate under a variety of conditions. Newer plasma tools have more adjustable pa-

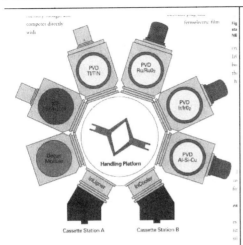

A cluster tool like this has a central load lock which shuffles wafers into different plasma reactors for etching, deposition, or stripping. Source: BPS

This footprint is too large to be economical.

This footprint is unbearable.

rameters, such as magnetic field shape and independent power sources, to make them more versatile.

- **Reliability.** In a factory, equipment failures cause expensive delays. Simple design can lead to more reliable plasma sources.

- **Small footprint.** Compactness is an important attribute when hundreds of machines need to be housed in a fabrication facility.

- **Benign materials.** To keep contamination down, very few materials are admissible in a plasma source. Since the wafer is silicon, Si walls are desirable. Often, glass or quartz, which are mostly Si, are used. Aluminum and alumina are common wall materials. Plasma sources which require internal electrodes would introduce metals into the chamber.

2. Elements of a plasma source

There are four main subsystems to a plasma source: the vacuum system, the gas handling system, the cooling system, and the discharge power source. Plasmas that require a magnetic field would also need field coils and their power supply.

Vacuum system

To make a plasma, we must first create a vacuum. Atmospheric pressure is 760 Torr, and operating pressure in a plasma reactor is generally between 1 mTorr and 100 Torr. The base pressure, before the chamber is filled with gas, has to be much lower than the operating pressure in order to keep down the partial pressure of contaminants. Thus, base pressures are at least 10^{-5}, and sometimes 10^{-6} (the Torr is understood). This is not usually a problem. Ultra-high-vacuum (UHV) systems can get down to 10^{-10}, but these are not generally needed. (High-energy accelerators can get down to below 10^{-20} Torr, approaching the vacuum of outer space.) The turbomolecular pump, or *turbopump*, is universally used nowadays. This has a multi-slotted fan blade that spins at a high velocity, physically blowing the gas out of the vacuum chamber. The rotor has to be supported by a very good bearing, sometimes oil cooled, or by magnetic suspension. The speed is controlled by an electronic circuit. Old pumps used oil or mercury vapor, which can get back into the chamber and contaminate it; but turbopumps are basically clean. The fan blade, however, cannot maintain the large pressure differential between high vacuum and atmospheric pressure; the air on one side would give so much drag that the blade could not spin at the required speed. So a turbopump has to be backed up by a *forepump*, or *backing pump*. There are many types of these, but they are all mechanical. For instance, a diaphragm pump moves a diaphragm back and forth and valves open and close to move the air from one side to the other. To pump the corrosive gases used in plasma processing, all the materials have to be chemically inert, and these dry pumps are considerably more expensive. The forepump generally provides a pressure, called the *forepressure*, of 1 to 50 mTorr, and the turbopump can then maintain the differential between this pressure and the base pressure.

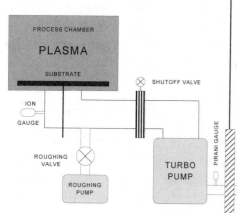

Because of their noise and exhaust, the forepump and roughing pump are usually put behind a wall, outside the clean room.

Gases are consumed in plasma processing, and large pumps are necessary to maintain a large flow rate. To maintain high conductivity, the pump is connected to the plasma chamber through a short, large-diameter pipe. Between them there is usually a gate valve. The hose from the turbo pump to the forepump does not need to be so large and short, since it handles the gas flow at a much higher pressure. The noisy forepumps are usually located on the other side of a wall. To be able to keep the turbopump running while the chamber is let up to atmospheric pressure to make a change, it is useful to connect the chamber to a *roughing pump* through a valve. This can bring the chamber down to a pressure ($\approx$50 mTorr) at which it is safe to open the gate valve to the turbopump.

Gas handling system.

The mixture of gases to be used in a process is formed in a gas manifold, into which gases from different tanks are fed through flow regulators. All this is electronically controlled. The gas mixture is then put into the process chamber through a *showerhead*, which is a circular tube with many equally spaced holes in it that distribute the gas uniformly around the inside circumference of the chamber. The flow rate is measured in *sccm* (standard cubic centimeters per minute), which is the number of cc of gas at STP flowing through per minute. The pumping rate, or speed S, of a pump, however, is measured in *liters per second*, which is a measure of volume, not amount of gas. Except at very high pressures, S does not depend on the pressure, so the number of sccm that a pump can remove depends on the operating pressure. In processes that consume a lot of gas, the flow rate must be high in order to keep the neutral pressure low. This is desirable, for instance, to keep the wafer sheath collisionless so that the accelerated ions are not deflected, or to keep dust particles from forming. Therefore, large turbopumps, with apertures of, say, 12 inches, and pumping speeds in the thousands of liters per second, can be found on plasma reactors. The gas handling system, with numerous inputs from tanks of gases, flow meters and flow controls, and computer interface, can be a large part of the plasma system.

Cooling system.

One of the disadvantages of plasma processing is that a lot of heat is generated. Walls of the chamber are usually water-cooled. Antenna wires are made of copper tubes with water flowing through them. The most critical cooling requirement is imposed by the wafer, which has to be maintained at a given temperature for each process, and which tends to be heated severely by plasma bombardment. Helium is introduced to the back side of the wafer through holes in the *chuck* which holds it. This gas is made to flow under the wafer to keep it at a uniform temperature. It is not necessary to create a space for the helium to flow; the underside of the wafer is usually rough enough.

Discharge power system.

To ionize and heat a plasma, electrical power is applied either at a radiofrequency (RF) or at a microwave frequency. The vast majority of sources use the industrially assigned frequency of 13.56 MHz. Some work at a harmonic or subharmonic of this, and some experimental sources run at frequencies higher or lower than this range. Electron cyclotron resonance (ECR) sources are driven at 2.45 GHz, the same as is used in microwave ovens.

RF sources are usually driven by a solid-state power amplifier with a built-in oscillator to generate the signal. The output into a 50-Ω cable is usually 2 kW or less. The cable then goes into a *matching network*, or *matchbox*, which performs the important function of transforming the impedance of the antenna-plasma system to the 50-Ω impedance of the rest of the circuit. Before passing through the matching network, the power goes through directional couplers which measure the power flowing into the antenna and back from it. This reflection has to be kept low (< 1%) to protect the amplifier and to make best use of its power. The main elements of the matching network are two (physically)

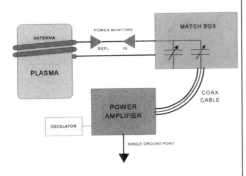

Elements of an RF power system. The frequency generator and power amplifier are usually in one chassis, while the matching circuit and the meters that measure the input and reflected power are in another chassis. Autotune circuits sense the amount of reflected power and automatically change the variable capacitors to minimize it.

large, adjustable vacuum capacitors. The tuning is done by varying the capacitances of these two elements. Since the RF current in the capacitors is displacement current in vacuum, there is very little power loss in such a circuit. Sometimes a variable inductor is used, consisting of two coils, one of which can be rotated to change the mutual coupling. Industrial tools invariably have *automatch* circuits, in which the tuning capacitors are automatically adjusted by motors driven by a circuit that detects the reflected power and tries to minimize it. Once the operating conditions of the plasma source are set, the automatch circuit has no problem finding the minimum and keeping the system tuned as the plasma conditions change. However, finding the vicinity of the correct match may be difficult initially. After the match circuit, the power is fed to the antenna through cables (several may be needed to carry the current) or a parallel transmission line. At this point there may be very high voltages, exceeding 1 kV. The length of the line affects the tuning conditions sensitively. In a capacitive discharge, the RF is connected directly to the internal electrodes. In an inductive discharge, the power goes to an external antenna, which is wound around the chamber in various ways depending on the type of source. In experimental systems there may be sensors to measure the RF voltage and current applied to the antenna.

ECR sources are driven by a magnetron providing 2.45-GHz power, which is transmitted in a waveguide. A "Magic T" device serves the function of the matching network in RF systems. The waveguide then goes to a horn antenna, which launches the microwave power into the plasma through a window. This vacuum window is a critical element, since it has to be made of a material such as quartz or ruby and tends to crack under high power. It also can be coated by deposits from the plasma. Since an ECR source has to strike a cyclotron resonance, magnetic coils have to provide the resonant field of 875 G somewhere in the plasma. Magnet coils are usually water-cooled copper tubes wound with many turns and held together by epoxy. They are driven by a low-voltage, high-current power supply such as those used for arc welding, only with better filtering.

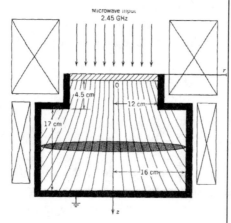

An ECR source, with the resonance zone shown shaded (from L & L).

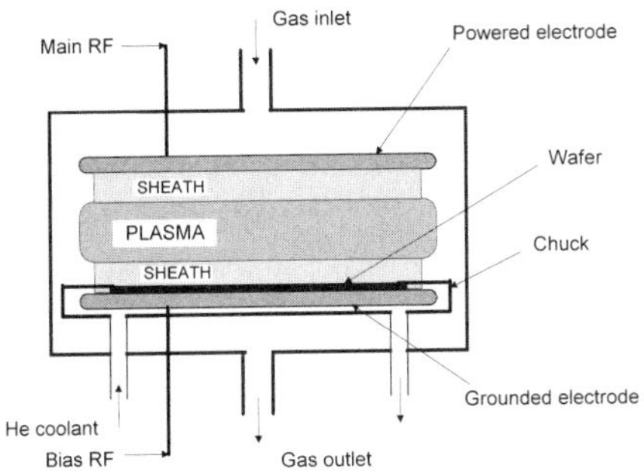

PRINCIPLES OF PLASMA PROCESSING
Course Notes: Prof. F.F. Chen

PART A4: PLASMA SOURCES II

V. RIE DISCHARGES (L & L, Chap 11, p. 327*ff*)

These simple devices, which were the staple of the industry until the mid-90s, consist of two flat, circular electrodes, about 20 cm in diameter, separated by about 10 cm. The wafer to be processed is mounted onto the bottom plate and held firmly by a "chuck", which includes connections for the helium coolant and for connecting to a *bias oscillator*, which we will discuss later. To produce the plasma, RF power may be applied to either or both plates. The sidewalls may be of an insulating material such as aluminum oxide, or a metal such as stainless steel, which can be grounded. For definiteness in what follows, we shall assume that the wafer-bearing plate is grounded and the upper plate oscillates at 13.56 MHz. Gas is fed into the vacuum chamber, and the RF field electric field causes the first few electrons (there are always a few from cosmic rays or whatever) to oscillate and gain enough energy to ionize atoms. The electrons thus freed will also gain energy and cause further ionizations. This *electron avalanche* quickly fills the chamber with plasma, whose density and temperature depend on the RF power applied and on the neutral pressure. The plasma is isolated from the electrodes and the walls by sheaths, and the RF fields are subsequently coupled to the plasma through the capacitances of the sheaths. These sheaths control the ion flux to the wafer, and it behooves us to examine them in some detail.

Fig. 1. Schematic of a parallel-plate capacitive discharge, called a Reactive Ion Etcher (RIE)

1. Debye sheath.

Consider first the sheath on a grounded wafer bounding a plasma that is *not* oscillating. Let the plasma potential (space potential) be V_s and the wafer potential be $V_w < V_s$. From our discussion of presheaths in Eq. (A1-22), the plasma density n_s at the sheath edge will be about $\frac{1}{2}n$. The ion flux through the sheath from the plasma to the wafer is given by

$$\Gamma_i = n_s c_s , \quad c_s = (KT_e / M)^{1/2} \tag{1}$$

The random flux of electrons entering the sheath is $n v_r$, where $v_r = \frac{1}{4}\bar{v}$, $\bar{v}$ being the average electron velocity in any direction (Chen, p. 228):

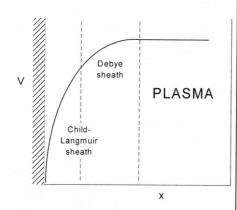

$$v_r = \frac{\bar{v}}{4} = \frac{1}{2}\left(\frac{2}{\pi}\frac{KT_e}{m}\right)^{1/2}. \qquad (2)$$

The flux of electrons getting through the sheath barrier to the wafer is then

$$\Gamma_e = nv_r\, e^{e(V_w - V_s)/KT_e}. \qquad (3)$$

Setting $\Gamma_i = \Gamma_e$ and solving for the *sheath drop*, we obtain

$$V_w - V_s = \frac{KT_e}{2e}\ln\left(\frac{\pi m}{2M}\right) \qquad (4)$$

This amounts to $-3.53T_{eV}$ for hydrogen and $-5.38T_{eV}$, **or about $5T_{eV}$,** for argon. The Debye length for $T_{eV} = 5$ and $n = 10^{11}$ cm^{-3}, say, is, from Eq. (A1-7),

$$\lambda_D = 7.4\sqrt{\frac{5}{0.1}} = 52\ \mu m.$$

The sheath thickness s can be obtained only by integration, but it is of order $5\lambda_D$; thus, in this case the Debye sheath is about 0.25 mm in thickness, and the sheath drop is about $5 \times 5 = 25$ V.

Fig. 2 Artificial separation of the sheath into a Debye sheath (which contains electrons) and a Child-Langmuir sheath (which has ions only).

2. Child-Langmuir sheath.

When a voltage is applied between the plates, the sheath drop cannot be 25 V on both plates; at least one of them must have a much larger sheath drop to take up the RF potential of hundred of volts that is applied. These large potential drops, much larger than KT_e, occur in a layer called a *Child-Langmuir sheath*, that joins smoothly onto the Debye sheath and extends all the way to the wall. This differs from the Debye sheath because only one charged species, in this case ions, exists in the C-L sheath, the electrons having almost all been turned back before they reach it. Those that remain are so few that they contribute a negligible amount to the charge in the C-L sheath. The current density j, voltage drop V_0, and thickness d are related by the Child-Langmuir Law of Space-Charge-Limited Diodes (Chen, p. 294, L & L, p. 165):

$$j = \frac{4}{9}\left(\frac{2e}{M}\right)^{1/2}\frac{\varepsilon_0 V_0^{3/2}}{d^2} \qquad (5)$$

We can equate this to the ion current density $j = en_s c_s$ and solve for d; the result is:

$$d^2 = \frac{4}{9}\frac{\varepsilon_0}{e}\left(\frac{2}{T_{eV}}\right)^{1/2}\frac{V_0^{3/2}}{n_s}. \qquad (6)$$

Multiplying and dividing by KT_e to form a factor equal to λ_D^2 [Eq. (A1-5)], we can express d in terms of λ_D as:

$$d = \frac{2}{3}\left(\frac{2V_0}{T_{eV}}\right)^{3/4}\lambda_D. \qquad (7)$$

This formula differs by $\sqrt{2}$ from standard treatments because we have evaluated λ_D in the plasma proper, not at the sheath edge, where the density is half as large. As an example, let $V_0 = 400$ V and $KT_e = 5$ eV; this gives $d = 15\lambda_D$, or about 0.8 mm for the example used above. Thus, the total sheath thickness $s + d$ is about $20\lambda_D = 1$ mm. This is much larger than feature sizes on the chip but much smaller than discharge dimensions. A density of 10^{11} cm^{-3} is high for an RIE plasma, however; total sheath thicknesses over 1 cm, an appreciable fraction of the discharge height, are often seen in RIE discharges at lower densities and higher temperatures. Note that d varies as $V_0^{3/4}$. This approximation is not really a good one, as the exact solution (Fig. 3) for the combined sheaths shows. The slope of 3/4 is followed only in very thick sheaths at very high potentials.

At the high pressures necessary to get high plasma densities, the collision mean free path of the ions can be shorter than the sheath thickness. Ions can then scatter in the sheath, thus making anisotropic etching more problematical.

3. Applying a DC bias

Consider a parallel-plate system with plate A (the wafer side) grounded. If plate B (the hot side) is also at zero potential, there will be identical sheaths of $\sim 5KT_e$ on each. For instance, if $KT_e = 2$ V, $V_s - V_B$ might be 10V. If plate B is now made more positive, so that $V_B = 5$V, say, $V_s - V_B$ must still be 10V, so V_s must rise to ≈ 15V. This is because if $V_s - V_B$ were only 5V, more electrons would flow to plate B than ions, and the loss of negative charge would immediately raise V_s. On the other hand, if plate B were to go negative, to -5V, say, V_s need not change. $V_s - V_B$ is now ≈ 15V, and the extra 5 volts is taken up by a Child-Langmuir sheath, while V_s is

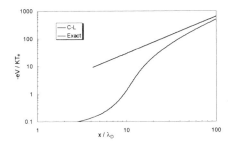

Fig. 3. An exact calculation for a plane sheath shows that C-L scaling is not followed unless the sheath is very thick (log-log scale).

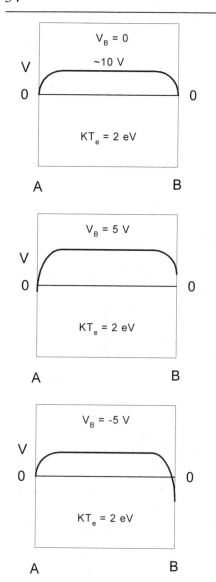

Fig. 4. Illustrating the change in plasma potential when one electrode is biased.

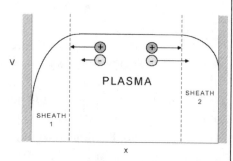

Fig. 5. Illustrating the slight difference in particle flows to asymmetric sheaths (from Part A1).

maintained at just below 10V by the sheath on plate A. Thus, *the plasma potential always follows the potential of the most positive electrode* or section of the wall. With an RF power supply driving plate B with a sinusoidal voltage, V_s will follow the *positive* excursions but will remain at the potential set by plate A during the *negative* excursions of plate B. Meanwhile, plate A (the wafer) will have a constant sheath drop (10V in our example) when plate B is negative, but will have a larger sheath drop with a C-L sheath whenever plate B is positive. Thus, the time-averaged sheath drop will be larger in the presence of an RF drive, and the average ion will impinge on the wafer with higher energy. Since the RF power controls the plasma density also, the ion current and energy for anisotropic etching cannot be controlled independently in a single-frequency RIE discharge.

To make this more quantitative and extend the treatment to asymmetric discharges, let the area of A be A_A and that of B be A_B. Thus, when these are equal, we have two similar plates at the top and bottom of the discharge, and when $A_B \ll A_A$, we have a small plate while the rest of the enclosure may be grounded. For the present, we do not consider a grounded sidewall, which would form a third electrode.

Using Eqs. (1) and (3), we can equate the ion and electron fluxes to both electrodes:

$$\Gamma_i = \tfrac{1}{2}nc_s , \qquad \Gamma_{er} = \tfrac{1}{4}n\overline{v} \qquad (8)$$

$$(A_A + A_B)\Gamma_i = \Gamma_{er}(A_A e^{e(V_A - V_s)/KT_e} + A_B e^{e(V_B - V_s)/KT_e}) \qquad (9)$$

where V_A and V_B are the potentials applied to the two electrodes. We can simplify this equation by setting $V_A = 0$ on the larger electrode and defining the following dimensionless quantities:

$$\varepsilon \equiv \frac{\tfrac{1}{2}nc_s}{\tfrac{1}{4}n\overline{v}} = \left(\frac{\pi m}{2M}\right)^{1/2}, \quad \eta \equiv \frac{eV}{KT_e}, \quad \delta \equiv \frac{A_B}{A_A} < 1. \quad (10)$$

Dividing by A_A, we obtain

$$(1 + \delta)\varepsilon = e^{-\eta_s} + \delta e^{(\eta_B - \eta_s)}.$$

The last term is valid only if $\eta_B < \eta_s$, since the electron flux cannot exceed the saturation value Γ_e. If $\eta_B > \eta_s$, $\exp(\eta_B - \eta_s)$ is replaced by 1. Thus, we have two cases: for $\eta_B < \eta_s$, we have

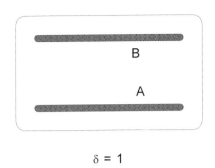

$\delta = 1$

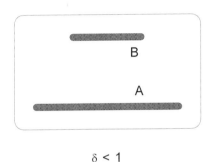

$\delta < 1$

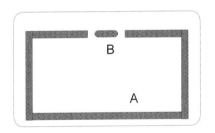

$\delta \ll 1$

Fig. 6. Capacitive discharge with asymmetric electrodes.

$$(1+\delta)\varepsilon = e^{-\eta_s}(1+\delta e^{\eta_B}), \quad \eta_s = \ln\left(\frac{1+\delta e^{\eta_B}}{(1+\delta)\varepsilon}\right) \quad (11)$$

and for $\eta_B > \eta_s$, we have

$$(1+\delta)\varepsilon = e^{-\eta_s} + \delta, \quad \eta_s = -\ln[(1+\delta)\varepsilon - \delta]. \quad (12)$$

Since the argument of the logarithm has to be positive, the latter case cannot occur unless

$$\varepsilon > \frac{\delta}{1+\delta} \approx \delta, \quad \delta < \varepsilon. \quad (13)$$

This means that B cannot draw saturation electron current unless the area ratio is less than ε, which is 0.0046 for argon. If A_B is that small (very unlikely), B is effectively a wall probe, and we shall use probe theory to describe it. For the normal case, Eq. (11) is valid, and V_s will follow positive excursions in V_B.

When no bias is applied ($V_B = 0$), Eq. (11) reduces to

$$\eta_s = \eta_{s0} = \frac{eV_s}{KT_e} = \ln\left(\frac{1}{\varepsilon}\right) = 5.38 \text{ for argon}, \quad (14)$$

as we saw following Eq. (4). This is the normal sheath drop. When $\delta = 1$, Eq. (11) becomes

$$\eta_s = \eta_{s0} + \ln[\tfrac{1}{2}(1 + e^{\eta_B})]. \quad (15)$$

As expected, this shows that the V_s is close to V_{so} when V_B is negative and approximately follows V_B when V_B is positive.

4. Applying an RF bias

One cannot apply a DC bias to a wafer, since at least some of the layers deposited on the wafer are insulating. However, it is possible to impose a time-averaged DC bias with RF. At RF frequencies, the ions are too massive to follow the fluctuations and will flow to each wall with the same flux Γ_i. The electrons respond so fast that they can maintain a Maxwellian distribution at every phase of the RF. Thus, the sheath at each phase of the RF will be the same as a DC sheath at the instantaneous voltage of the electrode. If we assume a sinusoidal oscillation of V_B with an amplitude $\hat{V}_B$,

$$V_B = \hat{V}_B \sin\omega t, \qquad \eta_B = \hat{\eta}_B \sin\omega t, \quad (16)$$

Eq. (11) gives the instantaneous plasma potential as

$$\eta_s = \ln\left[\frac{1 + \delta \exp(\hat{\eta}_B \sin \omega t)}{(1 + \delta)\varepsilon}\right]. \qquad (17)$$

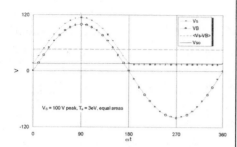

Fig. 7. The sheath drop and its dc average when the electrode voltage varies sinusoidally.

The behavior of V_s during one RF cycle is shown in Fig. 7 for $\delta = 1$ (equal areas) and $\hat{V}_B = 100$ V. As expected, the space potential rises with V_B when it goes positive, but remains around V_{s0} when V_B goes negative. The average space potential $\langle V_s \rangle$ is then higher than V_{s0}. Since $\langle V_B \rangle = \langle V_A \rangle = 0$, the average sheath drop on B is

$$\langle V_s - V_B \rangle = \langle V_s \rangle - \langle V_B \rangle = \langle V_s \rangle = \langle V_s - V_A \rangle \quad (18)$$

This is the accelerating potential seen by the ions and is the same on both electrodes. Thus, the ion energy impinging on the wafer is increased when RF power is applied to electrode B to strike the discharge. In this case the increase is from 16V to 46V. One can say that the sheath rectifies the RF, increasing the negative bias on the wafer relative to the plasma. In RIE discharges, the density of the plasma can be increased only by increasing the power applied, and this necessarily increases the ion energy. Though one cannot reduce the ion energy, one can increase it by applying another oscillator to the wafer electrode A. This is called an RF *bias oscillator*. It can be at the same frequency as the discharge power or at another (usually lower) frequency.

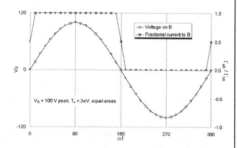

Fig. 8. The instantaneous electron current to the powered electrode.

We next consider the electron current to an electrode during an oscillation. The electron flux to B is

$$\Gamma_{eB} = A_B \Gamma_{er} \exp(\eta_B - \eta_s) \qquad (19)$$

The fractional current, normalized to the total ion current of Eq. (9), is then

$$\begin{aligned}
\frac{\Gamma_{eB}}{\Gamma_{tot}} &= \frac{A_B}{A_A + A_B}\frac{\Gamma_{er}}{\Gamma_i}e^{\eta_B}e^{-\eta_s} \\
&= \frac{\delta}{1+\delta}\frac{1}{\varepsilon}\left(\frac{(1+\delta)\varepsilon}{1+\delta e^{\eta_B}}\right)e^{\eta_B} = \left[1 + \delta^{-1}e^{-\hat{\eta}_B \sin \omega t}\right]^{-1}
\end{aligned}$$

$$\qquad (20)$$

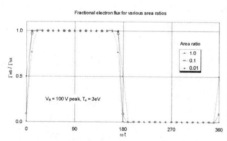

Fig. 9. The area ratio makes little difference in the current...

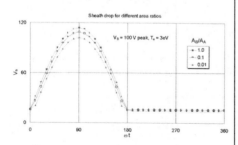

Fig. 10. ...or the sheath drop.

where we have used Eq. (17) for $\exp(-\eta_s)$. This is shown in Eq. (8) for $\delta = 1$. We see that the current flows only on the positive half-cycle of V_B but flows during that entire half cycle because we have set $V_B \gg KT_e/e$. This picture is not changed appreciably even if we made B a small electrode, as Figs. 9 and 11 show. Indeed, δ makes little difference in either the instantaneous sheath drop or the average sheath drop, as shown by calculations using Eqs. (17) and (18).

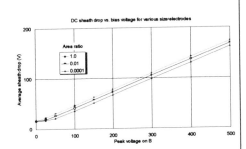

Fig. 11. Sheath drop vs. RF voltage
for various area ratios.

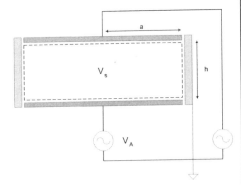

Fig. 12. A "triode" RIE reactor with
individually biased wafers and walls.

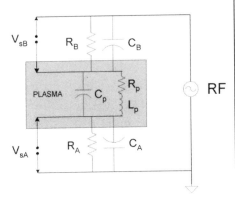

Fig. 13. Electrical representation of
the sheath-plasma system

A real RIE reactor, of course, has walls in addition to the top and bottom electrodes. If the sidewalls are insulators, no net current can flow to them, and therefore their surfaces must charge up to an oscillating voltage that follows V_s, keeping the sheath drop at V_{s0}. This oscillation may couple capacitively through the sidewalls to ground, in which case the walls are not completely insulating—they may be AC-grounded. If the sidewalls are grounded conductors, they must be treated as a third electrode. Now when different RF voltages and frequencies are applied to A and B, the total electron and ion fluxes must be set equal to each other when summed over all three conductors. The space potential will oscillate with both frequencies and the beat between them. The calculation is a trivial extension of what we have done here so far.

5. Displacement current

Up to now we have considered the plasma to be a perfect conductor, so that the sheath voltage V_s is the same on all boundaries. Actually, there is an appreciable resistivity to a low-temperature, weakly ionized plasma, and we must consider the plasma and the sheaths to be part of an electric circuit. The plasma can be represented by a resistance R_p and an inductance L_p in parallel with a capacitance C_p. R_p is due to electron collisions with neutrals and ions as they drift to carry the RF current. The effect of collisions with neutrals is given by the mobility formula given in Eq. (A2-15), and that with ions by Eq. (A2-6). L_p is due to the relative inertia of the ions and electrons, which causes them to respond differently to an AC field; this effect is negligibly small. C_p is the coupling from one plate to the other via displacment current; this is in parallel because that current would flow even if the plasma had infinite resistance. Since the plates are so far apart, this capacitance is also negligible. For the moment, let us assume $R_p = 0$ as before, so that we can concentrate on the displacement current. V_{sA} and V_{sB} are the sheath drops on each electrode, R_A and R_B the nonlinear conduction currents through the sheath, and C_A and C_B the sheath capacitances. Each sheath of area A has a capacitance

$$C = \varepsilon_0 A / d, \qquad d = s_{Debye} + d_{C-L}, \qquad (21)$$

and C will oscillate as d oscillates. The sheath impedance is $Z = 1/j\omega C$, and for a voltage V across the sheath, the displacement current I_d is given by

$$I_d = V/Z = j\omega CV = j\omega V \varepsilon_0 A / d. \qquad (22)$$

This RF current has to pass through both sheaths, so we have

$$I_d = j\omega\varepsilon_0\left(A_A V_{sA}/d_A\right) = j\omega\varepsilon_0\left(A_B V_{sB}/d_B\right), \qquad (23)$$

where V_{sA} is the sheath drop on sheath A, etc. Assuming the sheath thicknesses to be about the same on average, we have

$$\frac{V_{sB}}{V_{sA}} = \frac{A_A}{A_B}. \qquad (24)$$

Thus, there is a voltage divider action which depends on δ. In an asymmetric discharge, the smaller electrode sees more of the applied RF voltage than the larger electrode, so the average sheath drop is larger on the smaller electrode. This effect on the ion energy to the wafer is much larger than the δ-effect from sheath rectification, which is also in the opposite direction. To see the magnitude of the displacement current, we can compare it with the (electron) conduction current, which on average is of the order of the ion current $I_i = \frac{1}{2}neAc_s$. Thus,

$$\left|\frac{I_d}{I_i}\right| = \frac{2\pi f \varepsilon_0 V}{d}\frac{2}{n}\frac{1}{ec_s} \approx \frac{10^{12}}{n_{cm}}. \qquad (25)$$

Here we have taken $f = 13.56$ MHz, $V \approx 400$ V, $d \approx 1$ mm, and $KT_e \approx 5$ eV in argon. This ratio is unity when $n = 10^{12}$ cm^{-3}. In RIE tools, n is much smaller than this, so that the displacement current is dominant. The power from the RF source is coupled to the plasma through the sheath capacitances and drives the plasma current, carried by electrons, through the plasma, resistively heating and ionizing it. The sheath conduction currents discussed in the previous section play only a small rôle in the power balance, though they are responsible for the rectification effect that produces anisotropic ions.

A more rigorous treatment of the electrical characteristics of the RIE discharge would require a circuit analysis of the equivalent circuit including both the conduction and displacement currents through the sheath. The fact that the conduction currents are not sinusoidal would produce harmonics of the RF frequency. If a bias oscillator is of different frequency is imposed on the wafer, there will also be beat frequencies. To prevent the two oscillators from loading each other, a bandstop filter can be put on each RF supply to prevent the frequency of the other supply from reaching it. The displacement current can heat the wafer, increasing the cooling requirements, even though it does not help in giving the wafer a dc bias. If the bias power is large, it can heat the plasma and change its density and temperature.

6. Ion dynamics in the sheath

With the sheath drop V_s - V_w swinging positive every other half cycle, electron flux to the wafer will come in pulses. One might think that the ion flux will also come in pulses. This is not always true: it depends on the frequency. We consider two extremes. At high frequency, the ions move so slowly in response to the RF that they drift more or less smoothly in the average sheath field rather than the instantaneous one. To see this, we can calculate the ion motion. Let the sheath drop be represented by a sinusoidal voltage of 400V peak-to-peak, concentrated in a layer about 1 mm thick. The equation of motion of an ion is then

$$M(d\mathbf{v}/dt) = e\mathbf{E} = e\mathbf{E}_0 \cos \omega t . \qquad (26)$$

Here E_0 has the value $(200/10^{-3}) = 2 \times 10^5$ V/m. Integrating twice over a half cycle, we obtain

$$\mathbf{v} = \frac{d\mathbf{x}}{dt} = \frac{e\mathbf{E}_0}{M\omega} \sin \omega t,$$

$$\mathbf{x} = \frac{e\mathbf{E}_0}{M\omega^2} \left[-\cos \omega t \right]_0^\pi \quad \rightarrow \quad 1.3 \times 10^{-2} \text{ cm} \qquad (27)$$

Thus, the ions only jiggle by 0.1 mm at the RF frequency and are accelerated mainly by the time-averaged sheath field.

The ions do not all strike the wafer with the same energy, however, since they have been accelerated different amounts depending on the phase of the RF at which they enter and leave the sheath. Moreover, the ion energy distribution is *bi-modal*, meaning that it has two peaks with a valley in between. To see how this comes about, consider the second extreme in which the frequency is so low that sheath drop does not change while the ions traverse it. Let the discharge be symmetric, so that $\delta = 1$. Eq. (17) then can be written

$$e^{\eta_s} = \frac{1}{2\varepsilon} \left(1 + e^{\hat{\eta} \sin \omega t} \right), \qquad (28)$$

where $\hat{\eta}$ is the peak normalized potential applied to either

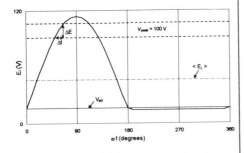

Fig. 14. Ions hitting the wafer have an energy depending on the sheath drop at the time. The number of ions at each energy depends on the time that sheath drop lasts; that is, on the reciprocal of the slope of this curve. At higher frequencies, ion transit time has to be taken into account.

electrode, the sheath drops being the same but opposite in phase on each electrode. Since we are neglecting the transit time, the ion energy E_i at any time is just the sheath drop $eV_s = \eta_s KT_e$ at that time. Assuming that the ions enter the sheath at a constant rate, the number of ions with energy E_i would be proportional to the time V_s remains at the corresponding value. Thus, on the graph of V_s vs. ωt (Fig. 7), which we reproduce in Fig. 14 on an expanded scale, the number of ions with energy between E_i and $E_i + \Delta E_i$ is proportional to $\Delta t / \Delta V_s$, or inversely proportional to the slope of the curve $V_s(\omega t)$.

To simplify the writing, we introduce the abbreviations

$$E \equiv E_i / KT_e = \eta_s , \qquad \theta \equiv \omega t , \qquad a \equiv \hat{\eta} . \quad (29)$$

Now we can write

$$f(E) = A \frac{d(\omega t)}{dE} = \frac{A}{dE / d\theta} , \quad (30)$$

where A is a normalization constant. Eq. (28) can be written in these variables and differentiated:

$$e^E = \frac{1}{2\varepsilon}\left(1 + e^{a\sin\theta}\right)$$

$$e^E dE = \frac{1}{2\varepsilon} e^{a\sin\theta} a\cos\theta d\theta \quad (31)$$

Substituting for e^E, we have

$$dE = \frac{a e^{a\sin\theta} \cos\theta d\theta}{1 + e^{a\sin\theta}} = \frac{a\cos\theta d\theta}{1 + e^{-a\sin\theta}} . \quad (32)$$

Hence,

$$f(E) = A \frac{1 + e^{-a\sin\theta}}{a\cos\theta} , \qquad \theta = \theta(E) . \quad (33)$$

From Eq. (31), we find

$$2\varepsilon e^E - 1 = e^{a\sin\theta} \quad \therefore \quad a\sin\theta = \ln(2\varepsilon e^E - 1) . \quad (34)$$

Since the argument of the logarithm must be ≥ 0, there is a minimum value of E given by

$$e^E > 1/2\varepsilon , \quad E_{min} = \ln(1/2\varepsilon) , \quad E_{i,min} = KT_e \ln(1/2\varepsilon) . \quad (35)$$

The denominator of Eq. (33) can be found as a function of E by using Eq. (34):

$$\cos^2\theta = 1 - \sin^2\theta = 1 - a^{-2}[\ln(2\varepsilon e^E - 1)]^2. \quad (36)$$

Hence, finally,

$$f(E) = A\frac{1 + \dfrac{1}{2\varepsilon e^E - 1}}{\left\{a^2 - [\ln(2\varepsilon e^E - 1)]^2\right\}^{1/2}} \quad (37)$$

$$= A\frac{2\varepsilon e^E}{(2\varepsilon e^E - 1)\left\{a^2 - [\ln(2\varepsilon e^E - 1)]^2\right\}^{1/2}}$$

There is also a maximum value of E given by the vanishing of the term in curly brackets:

$$\ln(2\varepsilon e^E - 1) = a, \qquad 2\varepsilon e^E = e^a + 1$$

$$E_{max} = \ln\left(\frac{e^a + 1}{2\varepsilon}\right). \quad (38)$$

It would be difficult to integrate Eq. (37) to find the normalization constant A, but there is an easier way. From Eq. (30), we have

$$1 = \int f(E)dE = \int A\frac{d\theta}{dE}dE = A\int_{-\frac{\pi}{2}}^{\frac{\pi}{2}} d\theta = A\pi. \quad (39)$$

Hence $A = 1/\pi$, and the ion distribution function at the wafer is

$$f(E) = \frac{1}{\pi}\frac{2\varepsilon e^E/(2\varepsilon e^E - 1)}{\left\{\hat{\eta}^2 - [\ln(2\varepsilon e^E - 1)]^2\right\}^{1/2}}, \quad (40)$$

where $E = E_i/KT_e$, $\hat{\eta}$ is the peak RF drive voltage normalized to KT_e, and ε is essentially the square root of the mass ratio.

The behavior of $f(E)$ can be seen as follows. This function diverges at E_{min} and E_{max}. For small E near E_{min}, $(2\varepsilon e^E - 1)$ is small, and $f(E)$ varies as $1/(2\varepsilon e^E - 1)$. For E near E_{max}, we may neglect the "1"s and $\ln(2\varepsilon)$. Then

$$f(E) \propto \frac{1}{\sqrt{\hat{\eta}^2 - E^2}} = \frac{1}{\sqrt{1 - (E_i^2/\hat{V}^2)}}.$$

The actual $f(E)$ will be smoother than the simplified one shown here, since the ions enter the sheath with a finite temperature and will also oscillate during their transit through a thick sheath. One would expect that the

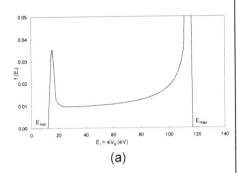

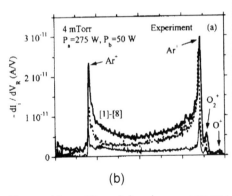

Fig. 15. Bimodal ion energy distribution (a) calculated from a simple model, and (b) measured experimentally. [Mizutani et. al., JVSTA **19**, 1298 (2001)]

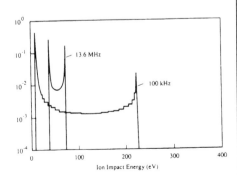

Fig. 16. The IEDF is narrower if the RF sheath oscillates during the ions' transit time.

ions would have the energy of the DC sheath drop, spread out by the sheath oscillation during their last jiggle. Thus, $f(E)$ will be relatively narrower but still bimodal. The ion flux that provides directionality to the etching process is therefore not completely controllable; the ions cannot all have the same energy. In laboratory experiments it has been shown that changing the bias oscillator's waveform to a sawtooth-like wave can produce a more uniform $f(E)$, but this has not been shown to be beneficial in practice.

To get an idea of when the analytic solution of Eq. (40) is valid, consider the ion excursion calculated in Eq. (27). If we set x equal to the sheath thickness d and solve for ω, we obtain $\omega \approx 3 \times 10^8$ sec^{-1}, or $f \approx 50$ MHz. Thus, unless $f \ll 50$ MHz, the spread in E_i will be smaller than what we have computed neglecting the ion transit time. This has been confirmed by both computation and experiment.

7. Other effects in RIE reactors

Stochastic heating. In most discharges, including the RIE, the plasma is maintained by ohmic heating; that is, the electron current is driven through the plasma by the applied electric field, and the electrons gain energy from the electric field between collisions. As they collide, the electron energy distribution function $f_e(E)$ (called *EEDF* in this industry) rapidly thermalizes into a Maxwellian but with a slight excess of high-energy electrons. This distortion of the EEDF is due to the fact that high energy electrons have a smaller collision cross section, as we have seen in Part A2. With or without this distortion, the small number of high-energy electrons in the "tail" of the EEDF is responsible for ionizing collisions that maintain the population of charged particles against diffusion to the walls. If it were not for the collisions, the electrons would simply oscillate back and forth as the RF field changed sign and lose as much energy as they gained. The collisions scatter the electrons so that they are moving in a different direction when the RF field reverses, and therefore do not lose exactly the same amount of energy that they gained. This scattering is responsible for the conversion of RF energy into plasma energy and the for the resistivity of the plasma and its ohmic heating.

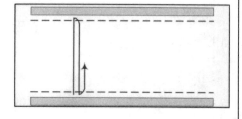

Fig. 17 Collisionless sheath heating mechanism

There is another randomizing mechanism besides collisions with ions or neutrals, however. Electrons can collide with sheaths. Most electrons are reflected by the sheath field when they approach a wall or electrode, gaining as much energy on reflection as they lost upon incidence. If the sheath field changes in the meantime, however, the gain and loss may not cancel out. Some electrons, namely those that enter the sheath when the field is low and leave the sheath when the field is high, will have a net gain in energy. Others will have a net loss, but those that gain energy are important because they contribute to the tail of the Maxwellian. A few electrons will have such a velocity that, as they travel back and forth between the electrodes, they will arrive at each sheath just at the right phase and experience repeated accelerations. Such collisionless sheath heating has been treated extensively in theory and computation, but there has been little experimental evidence of its importance. Perhaps this effect contributes to the fact that RIE reactors tend to have higher electron temperatures than other tools.

High pressure discharges

At the opposite extreme, RIE discharges can be run at high pressure (100–500 mTorr), where the collision mean free path is so short that many collisions occur within the sheath. Such reactors are used, for instance, in some SiO_2 etchers and in very large-area deposition machines for coating amorphous silicon onto glass substrates for flat-panel displays. The gap in such devices can be only 1–2 cm, and the sheaths can occupy most of the gap, leaving only a small quasineutral region. The EEDF there has been found to contain a population of very low energy electrons—less than 1 eV. The electron heating mechanism is quite different from that in low-pressure discharges. Most of the electrons gain energy in the sheaths, where the electric fields can be large. By the time they diffuse to the quasineutral region, they have lost most of their energy by colliding with neutrals and ions, giving rise to the low-T_e component.

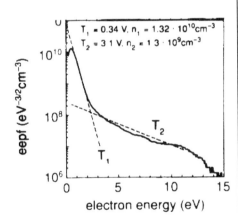

Fig. 18. A low-T_e bi-Maxwellian EEDF found in high-pressure RIEs [Godyak et al., Phys. Rev. Lett. **65**, 996 (1990)].

Gas flow and dust formation

Since capacitive discharges are less efficient ionizers than other plasma sources that we shall discuss, higher pressure is needed to obtain a given plasma density. This is fine for deposition but is deleterious for etching, because collisions in the sheath can spoil the anisotropy. In deposition especially, a large amount of gas is consumed, so that the neutral gas must flow through the chamber at a high rate. The collisions of the gas with ions will push the ion population in the direction of the flow. This effect, which is negligible in low-pressure discharges, must be taken into account in calculating the steady-state equilibrium of RIE discharges. This fact is dramatically illustrated by the observation of dust particles.

The nucleation of solid particles in a processing plasma from molecules produced in etching or deposition processes occurs by a mechanism that is not yet completely understood. These particles start with submicron, perhaps nanometer, sizes and grow to diameters of 1 micron or even greater. When they reach macroscopic sizes, dust particles can be seen by sweeping a laser beam across them. A sheath builds up on each particle larger than λ_D to equalize the ion and electron fluxes to it, and the particle thus becomes negatively charged, containing the order of 10^4 electron charges. This charge Q can easily be computed, since the capacitance of the particle is given by CV = Q, where V is the same sheath drop that we calculated previously for electrodes and walls. Since the particles are negative, they will accumulate at local maxima of the potential, if there are any. Otherwise, they can find a stable position above the wafer where the repulsion of the sheath field is balanced by the downward pressure exerted by the gas flow. By making grooves in the substrate or inserting dielectric or other materials under it, it is possible to create a path of maximum potential which will lead the dust out of the discharge, but these methods are probably not practical in large-scale production. When the discharge is turned off, the negative dust is drawn onto the wafer by the collapsing sheath field, thus destroying any circuits that they land on. Note that putting the wafer on the upper electrode will not help, since gravity is much weaker than the electric fields involved. Particulate generation in RIE discharges is a serious problem that lowers the yield of viable chips on a wafer. This problem is not as serious in low-pressure discharges, where dust formation is much less likely.

The GEC reference cell

RIE discharges were developed by trial and error without a good understanding of their operation. When RIE tools were dominant in the industry before the 1990s, many experiments were made to elucidate the mechanisms in these sources such as those we have discussed here. However, each machine was different, and it was difficult to draw any universal conclusions. The participants in the Gaseous Electronics Conference (GEC) at which many of these results were presented decided to design a standard configuration that everyone

Fig. 19. Dust particles suspended above three wafers and illuminated by a scanning laser beam [G.S. Selwyn, Plasma Sources Sci. Technol. **3**, 340 (1994)].

could duplicate for their experiments. Thus, the dimensions, materials, and mechanical and electrical design of this small machine (for processing 4" wafers) were specified. This *GEC reference cell* served to make observations of one group relevant to those of other groups. When inductive plasma sources were introduced, however, they differed from one another qualitatively, and, moreover, their designs were closely guarded secrets. There is no standard configuration for the more advanced plasma tools, and their development is pursued independently by each company.

8. Disadvantages of RIE reactors

Since the RF power controls both the plasma density and the oscillations in V_s, the ion flux to the wafer cannot be varied independently of the ion energy. Capacitive discharges tend to have larger RF electric fields, and hence the minimum ion energy is relatively high, as is the spread in ion energies. Being less efficient in ionization than inductive discharges, RIE discharges produce lower plasma densities and require higher operating pressures. It is then more difficult to avoid ion collisions in the sheath and the formation of particulates. The electron temperature tends to be higher than in inductive discharges, and this could lead to a larger heat load on the wafer as well as other deleterious effects such as oxide damage. There is no obvious way to control the EEDF so as to adjust the populations of the chemical radicals at the wafer level. In spite of these disadvantages, RIE reactors are still best for certain processes such as PECVD (plasma enhanced chemical vapor deposition) where feed gas can be distributed evenly over a large area by covering one electrode with holes. Contrary to common perception, some RIE discharges are found to cause *less* oxide damage than inductive devices. These "dinosaurs" are still alive and still incompletely understood. And they have the advantage of being small and simple.

9. Modified RIE devices

In addition to applying different RF powers and frequencies to the two electrodes and introducing a ground electrode, more substantive changes have been made to RIE reactors to improve their performance. The most popular of these is the MERIE: Magnetically Enhanced RIE. A magnetic field **B** is applied to the discharge parallel to the wafer surface. This reduces the diffusion losses of plasma in the directions perpendicular to **B** and enhances the density. The flux of electrons into the sheath is also reduced, requiring a smaller Coulomb barrier, and thus reducing the fluctuation amplitude of the sheath voltage. However, it has been found that such a B-field increases damage to thin oxide layers, an effect attributed to accumulation of charge via the $\mathbf{E} \times \mathbf{B}$ drifts of the electron guiding centers. The plasma as a whole can become nonuniform because of these drifts. To avoid this, the direction of the magnetic field can rotated slowly in the azimuthal direction by using two sets of field coils separated by 90° and driven 90° out of phase.

Another modification is to insert, half-way between the electrodes, a grounded plate with many small holes drilled in it. The discharge then tends to break up into numerous small discharges, one in each hole. These discharges are intense and can produce a high density even after diffusing into a uniform plasma at the wafer level. The plate also serves to isolate the RF fields produced by the primary RF drive and the bias oscillator. This device is confusingly called a *hollow cathode discharge*, presumably because real hollow cathode discharges have a source of electrons on axis and a ring or cylindrical anode (the positive electrode in a DC discharge) surrounding it.

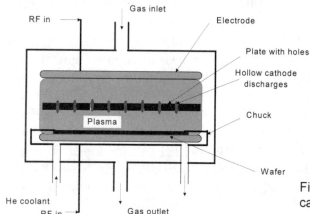

Fig. 20. A "hollow cathode" parallel-plate capacitive discharge

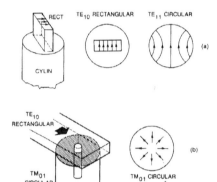

Fig. 1. Mechanism of electron cyclotron heating

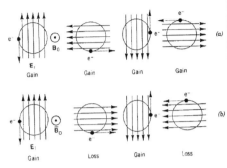

Fig. 2. Various microwave mode patterns and ways to couple to them.

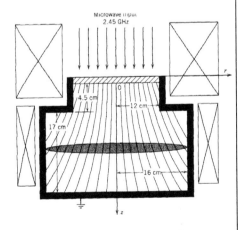

Fig. 3. The shaded region is the resonance zone (L & L, p. 429).

PRINCIPLES OF PLASMA PROCESSING
Course Notes: Prof. F.F. Chen

PART A5: PLASMA SOURCES III

VI. ECR SOURCES (L & L, Chap. 13, p. 412)

ECR discharges require a magnetic field such that the electrons' cyclotron frequency is in resonance with the applied microwave frequency, usually 2.45 GHz. Both the large magnetic field of 875G and the microwave waveguide plumbing make these reactors more complicated and expensive than RIE reactors. Unless one uses tricky methods that depend on nonuniform magnetic fields and densities, microwaves cannot penetrate into a plasma if $\omega_p > \omega$. At 2.45 GHz, that means that the maximum density that can be produced, in principle, is $100 \times (2.45/9)^2 = 7.4 \times 10^{10}$ cm^{-3} [Eq. (A1-9)]. However, this does not hold in the near-field of the launching device, usually a horn antenna or a loop or slot coupler. Densities of order 10^{12} cm^{-3} have been produced in ECR reactors because the free-space wavelength of 2.45-GHz radiation is 12.2 cm, and the interior of a 10 cm diam plasma is still within the near-field. This is discussed in more detail later.

In cyclotron heating, electrons gyrate around the B-field at ω_c; and if the microwave field also rotates at this frequency, an electron will be pushed forward continuously, gaining energy rapidly. Those electrons moving in the wrong direction will be decelerated by the field, but will eventually be turned around and be accelerated in phase with the field. Though resonant electrons gain energy only in their cyclotron motion perpendicular to **B**, they collide rapidly with other electrons and, first, become isotropic in their velocity distribution and, second, transfer their energy to heat the entire electron population. Since a thermal electron can lose only its small thermal energy while an electron in the right phase can gain 100s of eV of energy while it is in resonance, there is a net gain of energy by the distribution as a whole. At very low pressures, electrons can gain 1000s of eV and become dangerous, generating harmful X-rays when they strike the wall. Fortunately, there are two mitigating factors besides collisions: electrons do not stay in the resonance zone very long because of their thermal velocities, and microwave sources tend to be incoherent, putting out short bursts of radiation with changing phase instead of one continuous wave.

Actual ECR reactors in production have nonuniform

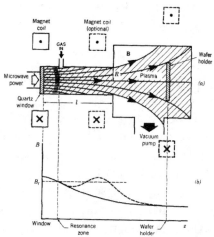

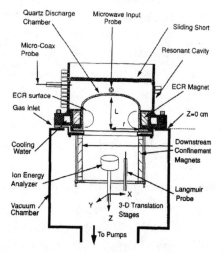

Fig. 4. In this case, a second coil can produce more resonance zones.

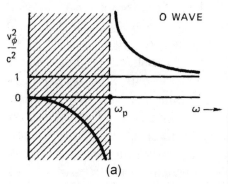

Fig. 5. This experimental device has an annular resonance zone.

O WAVE

(a)

Fig. 6. Dispersion curves for (a) the O-wave, (b) the L and R waves, and (c) the X-wave.

magnetic fields, so that **B** cannot be 875G everywhere. In a diverging magnetic field, there are resonance zones, usually shaped like a shallow dish, in which the field is at the resonance value. Electrons are heated when they pass through this zone, and the time spent there determines the amount of heating. Note that the microwave field does not have to be circularly polarized. A plane-polarized wave can be decomposed into the sum of a right- and a left-hand polarized wave. An electron sees the left-polarized component as a field at $2\omega_c$ and is not heated by it; the right-hand component does the heating.

A microwave signal at exactly ω_c will not travel, or propagate, through the plasma. This is because electromagnetic (e.m.) waves are affected by the charges and currents in the plasma and therefore behave differently than in vacuum or in a solid dielectric. The relation between the frequency of a wave and its wavelength λ is call its *dispersion relation*. For an e.m. wave such as light, microwave, or laser beam, the dispersion relation is

$$c^2 k^2 = \omega^2 - \omega_p^2$$

or

$$\frac{\varepsilon}{\varepsilon_0} = \frac{c^2}{v_\phi^2} = \frac{c^2 k^2}{\omega^2} = 1 - \frac{\omega_p^2}{\omega^2}, \qquad (1)$$

where $k \equiv 2\pi / \lambda$. In Part A1 we have already encountered the dispersion relations for two *electrostatic* waves: the plasma oscillation ($\omega = \omega_p$, which does not depend on k) and the ion acoustic wave ($\omega = k c_s$, which is linear in k). The dispersion curve for Eq. (1) is shown in Fig. 6(a) [from Chen, p. 128]. We see that if $\omega < \omega_p$ (the shaded region), the wave cannot propagate. Its amplitude falls exponentially away from the exciter and is confined to a skin depth. It is *evanescent*. We will need this concept later for RF sources. This behavior is specified by Eq. (1), since k^2 becomes negative for $\omega < \omega_p$. Note that for $\omega > \omega_p$ the waves travel faster than light! This is OK, since the group velocity will be less than c, and that's what counts.

The dispersion curve becomes more complicated when there is a dc magnetic field $\mathbf{B}_0$. For waves traveling along $\mathbf{B}_0$, the dispersion relation is

$$\frac{c^2 k^2}{\omega^2} = 1 - \frac{\omega_p^2}{\omega^2} \frac{1}{1 \mp \dfrac{\omega_c}{\omega}}. \qquad (2)$$

Here we have two cases: the upper sign is for right-hand

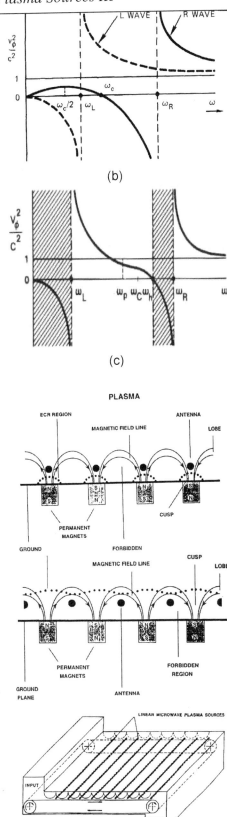

(b)

(c)

Fig. 7. A slotted-waveguide ECR source.

circularly polarized (R) waves, and the bottom one for left (L) polarization. These are shown in (b). Of interest is the R-wave, which can resonate with the electrons (the solid line). At ω_c, the curve crosses the axis; the phase velocity is zero. As the wave comes in from the low-density edge of the plasma, it sees an increasing n, and the point marked by ω_R shifts to the right. For convenience consider the diagram to be fixed and the wave to come in from the right, which is equivalent. After the wave passes ω_R, it goes into a forbidden zone where k^2 is negative. The resonance zone is *inaccessible* from the outside. The problem is still there if the wave propagates *across* B_0 into the plasma. This is now called an extraordinary wave, or X-wave, whose diagram is shown in (c). Here, also, the wave encounters a forbidden zone before it can reach cyclotron resonance.

Fortunately, there are tricks to avoid the problem of inaccessibility: shaping the magnetic field, making "magnetic beaches", and so on. ECR reactors can operate in the near-field; that is, within the skin depth, so that the field can penetrate far enough. Though ECR sources are not in the majority, they are used for special applications, such as oxide etching. There are many other industrial applications, however. For instance, there are ECR sources made for diamond deposition which leak microwaves through slots in waveguides and create multiple resonance zones with permanent magnet arrays, as shown in Fig. 7. By choosing magnets of different strengths, the resonance zone can be moved up or down. These linear ECR sources can be arrayed to cover a large area; as in Fig. 7 and 8.

There are also microwave sources which use surface excitation instead of cyclotron resonance. For instance, the so-called "surfatron" sources shown in Fig. 9 has an annular cavity with a plunger than can be moved to tune the cavity to resonance. The microwaves are then leaked into the cylindrical plasma column to ionize it. The plasma is then directed axially onto a substrate.

VII. INDUCTIVELY COUPLED PLASMAS (ICPs)

1. Overview of ICPs

Inductively Coupled Plasmas are so called because the RF electric field is *induced* in the plasma by an external antenna. ICPs have two main advantages: 1) no

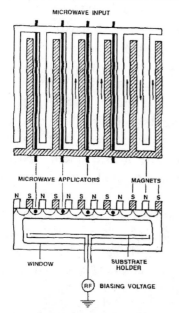

Fig. 8. A large-area ECR source

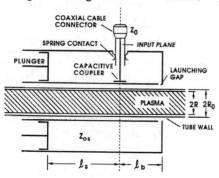

Fig. 9. A "surfatron" ECR source

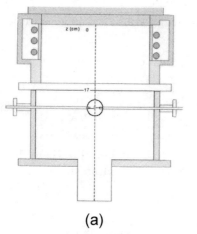

(a)

Fig. 11. A commercial ICP mounted on top of an experimental chamber (Plasma-Therm)

internal electrodes are needed as in capacitively coupled systems, and 2) no dc magnetic field is required as in ECR reactors. These benefits make ICPs probably the most common of plasma tools. These devices come in many different configurations, categorized in Fig. 10.

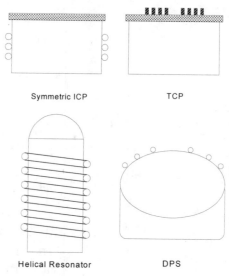

Fig. 10. Different types of ICPs.

In the simplest form, the antenna consists of one or several turns of water-cooled tubing wrapped around a ceramic cylinder, which forms the sidewall of the plasma chamber. Fig. 2 shows two commercial reactor of this type. The spiral coil acts like an electromagnet, creating an RF magnetic field inside the chamber. This field, in turn generates an RF electric field by Faraday's Law:

$$\nabla \times \mathbf{E} = -d\mathbf{B}/dt \equiv \dot{\mathbf{B}}, \qquad (3)$$

B-dot being a term we will use to refer to the RF magnetic field. This field is perpendicular to the antenna current, but the E-field is more or less parallel to the antenna current and opposite to it. Thus, with a slinky-shaped antenna, the E-field in the plasma would be in the azimuthal direction.

Rather than apply the RF current at one end of the antenna and take it out at the other, one can design a *helical resonator*, which is a coil with an electrical length that resonates with the drive frequency. The antenna then is a tank circuit, and applying the RF at one point will make the current oscillate back and forth from one end of the antenna to the other at its natural frequency. Such a resonant ICP is shown in Fig. 13.

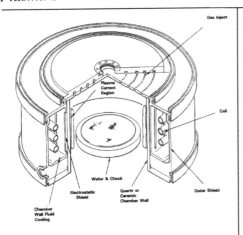

Fig. 12. A similar ICP by Prototech.

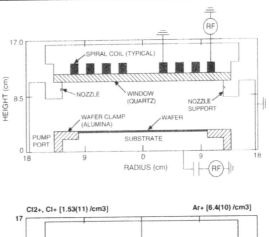

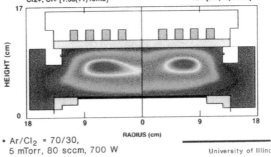

Fig. 14. Diagram of a TCP and a simulation by M. Kushner (U. Illinois, Urbana).

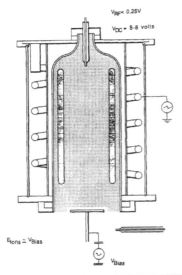

Fig. 13. Diagram and photo of a helical resonator (Prototech).

Another configuration called the TCP (Transformer Coupled Plasma) uses a top-mounted antenna in the shape of a flat coil, like the heating element on an electric stove. This design is meant to put RF energy into the center of the plasma, near the axis. Fig. 14 shows a diagram of a Lam Research TCP together with a computer simulation of its plasma. This will be discussed further later.

If one combines an azimuthal winding with a spiral winding, the result is a dome-shaped antenna, which is used by Applied Materials in their DPS (Detached Plasma Source) reactors (Figs. 15, 16). These sources are farther removed from the wafers, allowing the plasma to diffuse and become more uniform.

In addition to inductive coupling, there can also be capacitive coupling, since a voltage must be applied at least to one end of the antenna to drive the RF current through it. Since this voltage is not uniformly distributed, as it is in a plane-parallel capacitive discharge, it can cause an asymmetry in the plasma density. On the other hand, this voltage can help to break down the plasma, creating enough density for inductive coupling to take hold.

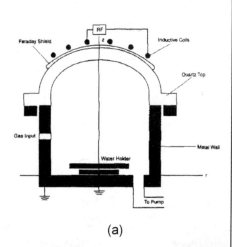

(a)

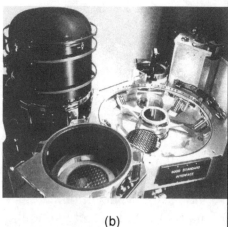

(b)

Fig. 15. A DPS ICP reactor (Applied Materials).

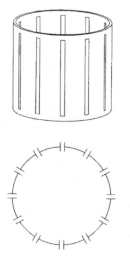

Fig. 17. Diagram of a Faraday shield

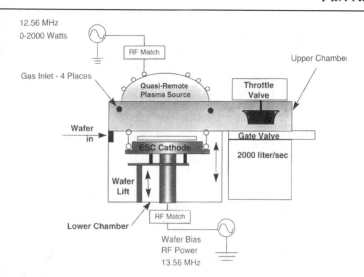

Fig. 16. Parts of a DPS reactor (Applied Materials).

Some machines shield out the capacitive coupling by inserting a *Faraday shield* between the antenna and the chamber. Such a shield is simply a thin sheet of metal which can be grounded. However, slits must be cut in the shield to allow the inductive field to penetrate it. These slits are perpendicular to the direction of the current flow in the antenna; then induced currents in the shield that would otherwise create a B-dot field opposing that of the antenna would be unable to flow without jumping across a gap. Such a shield is particularly important in helical resonators, since the standing wave in the antenna can cause very large voltages to develop. A Faraday shield can be seen in Fig. 13, and a diagram of it is shown in Fig. 17. Another essential feature is the electrostatic chuck, or ESC, which can be seen in Fig. 15. This uses electric charges to hold the wafer flat and will be discussed later.

Antennas do not present much resistance to the power supply, but when the plasma is created, the RF energy is absorbed, and this appears as a resistance to the power supply. Currents in the plasma induce a *back-emf* into the antenna circuit, making the power supply work harder, and this extra work appears as plasma heat. The part of the back-emf that is in phase with the antenna current appears as an added antenna resistance, and the part that is 90° out of phase appears as added or reduced inductance. Normally, the plasma density is low enough that the currents in the plasma are too small to affect the inductance much. Since most power supplies must be matched to a 50-Ω load, a *matchbox* or *tuning circuit* is needed to transform the impedance of the antenna to 50Ω, even under changing plasma loads. This will be discussed quantitatively later.

2. Normal skin depth

Just as a plasma can distribute its charges so as to shield out an applied voltage, it can also generate currents to shield out an applied magnetic field. In Eq. (1) we saw that electromagnetic waves cannot propagate through a plasma if $\omega < \omega_p$. More precisely, let the e.m. wave vary as $\cos[i(kx - \omega t)] = \text{Re}\{\exp[i(kx - \omega t)]\}$. (We normally use exponential notation, in which the "Re" is understood and is omitted.) If k is imaginary, as it is in ICPs, in which $\omega \ll \omega_p$, the wave will vary as $e^{-\text{Im}(k)x}\cos\omega t$ as it propagates in the x direction. The *evanescent* wave does not oscillate but decays exponentially upon entering the plasma. In the case of most ICPs, $f = 13.56$ MHz while ω_p is measured in GHz, so that $\omega \ll \omega_p$, and Eq. (1) can be approximated by $k^2 = -\omega_p^2/c^2$, and $\text{Im}(k) = \omega_p/c$. The wave then varies as $\exp(-x/\delta_s - i\omega t]$, where the characteristic decay length, called the *exponentiation* distance or *e-folding length*, is $1/\text{Im}(k)$. This decay length is defined as the *skin depth* δ_s:

$$\delta_s = \delta_c \equiv c / \omega_p . \qquad (4)$$

The quantity δ_c is called the collisionless skin depth and is the same as δ_s here because we have so far neglected collisions. Note that δ_c depends on $n^{-1/2}$; the current layer doing the shielding can be thinner if the plasma is dense. The diagram shows that as the current J in the antenna increases, a B-field is induced in the plasma, and this in turn drives a shielding current in the opposite direction. This current decays away from the wall with an e-folding distance d_s.

Using Eq. (A1-9) for f_p, we find that δ_c for a 10^{12} cm^{-3} plasma is of order 0.5 cm. We would expect that very little RF power will get past the skin layer and reach the center of the plasma, perhaps 10 cm away. This is the reason stove-top antennas were invented. However, it is found, amazingly, that even antennas of the first type in Fig. 10 can produce uniform densities across the whole diameter. This is one of the mysteries of RF sources and has given this field the reputation of being a "black art". A possible explanation will be given later.

ICPs, however, are not collisionless, and one might think that the skin depth would be increased by collisions. To account for collisions of frequency ν_c, we need merely replace Eq. (1) with

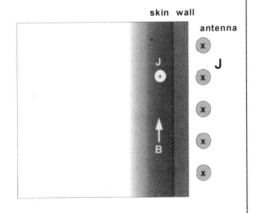

Fig. 18. Illustrating the currents in the antenna and in the skin layer.

$$\frac{c^2 k^2}{\omega^2} = 1 - \frac{\omega_p^2}{\omega(\omega + i\nu_c)} \ . \tag{5}$$

The complex denominator makes it hard to see how the skin depth is changed, but we can look for an approximation. Since $\omega \approx 9 \times 10^7 \ sec^{-1}$, $\nu_{en} = 3 \times 10^{13} \langle \sigma v \rangle_{en}$ p(mTorr), $\sigma \approx 5 \times 10^{-16} \ cm^2$, and $v \approx 10^8 \ cm/sec$ (for $KT_e \approx 4 \ eV$), we find that $\nu_c < \omega$ for $p < 60$ mTorr. If $\nu_c \ll \omega$, we can expand the denominator in a Taylor series, obtaining

$$c^2 k^2 \approx \omega^2 - \omega_p^2 \left(1 + \frac{i\nu_c}{\omega} \right)^{-1} \approx -\omega_p^2 \left(1 - \frac{i\nu_c}{\omega} \right).$$

Here we have taken $\omega_p^2 \gg \omega^2$, which is well satisfied. Taking the square root by Taylor expansion, we now have

$$k \approx \pm i \frac{\omega_p}{c} \left(1 - \frac{i}{2} \frac{\nu_c}{\omega} \right), \quad \delta_s = |\operatorname{Im}(k)|^{-1} \approx \delta_c \left[1 + O\left(\frac{\nu_c^2}{\omega^2} \right) \right].$$

We see that the real part of k has acquired a small value as a result of collisions, but the imaginary part is unchanged to first order in ν_c/ω unless our assumption of $\nu_c \ll \omega$ breaks down. At the mTorr pressures that ICPs operate at, the skin depth is quite well approximated by δ_c.

With a computer, it is easy to solve Eq. (5) using the collision data given in Part A2, and some results are shown in Fig. 19. These were computed for $f = 2$ MHz, used by some ICPs, rather than the more usual 13.56 MHz, in order to increase ν/ω and bring out the effect of collisions more clearly. We see in (a) that d_s decreases with $\sqrt{n}$ as expected, but it hits a lower limit imposed by collisions. In (b), we see that the increase of d_s with pressure depends on T_e, which affects $\nu_c = n_n \langle \sigma v \rangle$.

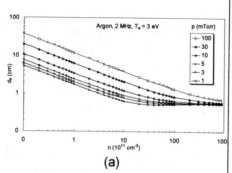

(a)

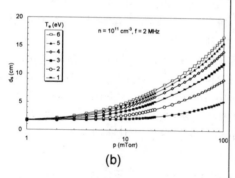

(b)

Fig. 19. (a) Classical skin depth vs. density at various neutral pressures. (b) Effect of argon pressure on skin depth at various electron temperatures.

3. Anomalous skin depth

In plasma physics, classical treatments like the above are often doomed to failure, since plasmas are tricky and more often than not are found experimentally to disobey the simple laws of electromagnetics. They can do this by deviating from strict Maxwellian distributions and generating internal currents and charges that are not included in simple formulations. However, in this case, observations show that Eq. (5) is correct . . . up to a point. Fig. 20 shows data taken in the ICP of Fig. 11 with a magnetic probe measuring the RF magnetic

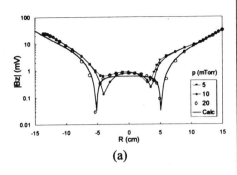

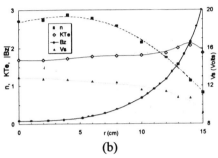

Fig. 20. (a) Decay of the RF field excited by a loop antenna. (b) Profiles of n (10^{11} cm^{-3}), KT_e (eV), plasma potential V_s (V), and B_z (arbitrary units) in an ICP discharge in 10 mTorr of argon, with P_{rf} = 300W at 2 MHz.

field B_z. In the outer 10 cm of the cylinder, the semilog plot shows $|B_z|$ decreasing exponentially away from the wall with a scalelength corresponding to d_s. (The exact solution here is a Bessel function because of the cylindrical geometry.) However, as the wave reaches the axis, it goes through a null and a phase reversal, as if it were a standing wave. This behavior is entirely unexpected of an evanescent wave, and such observations gave rise to many theoretical papers on what is called *anomalous skin depth*. Fast, ionizing electrons are created in the skin layer, where the E-field is strong. These velocities, however, are along the wall and do not shoot the electrons toward the interior. Most theorists speculate that thermal motions take these primary electrons inward and create the small reversed B-field there. This effect, however, should decrease with pressure, and Fig. 20a shows that the opposite is true. The explanation is still in dispute. In Fig. 20b, the exponential decay of B_z is seen on a linear scale. Also shown is the plasma density, which *peaks* near the axis. Since the RF energy, proportional to B_z^2, is concentrated near the wall, as the T_e profile confirms, it is puzzling why the density should peak at the center and not at the wall. We have explained this effect recently by tracing the path of an individual electron as it travels in and out of the skin layer over many periods of the RF. An example is shown in Fig. 21. The result depends on whether or not the Lorentz force $F_L = -e\mathbf{v} \times \mathbf{B}$ is included. This is a nonlinear term, since both $\mathbf{v}$ and $\mathbf{B}$ are small wave quantities. This force is in the radial direction and causes the electron to hit the wall at steeper angles. The electrons are assumed to reflect from the sheath at the wall. The steeper angles of incidence allow the fast electrons to go radially inwards rather than

Fig. 21. Monte Carlo calculation of a electron trajectory in an RF field with (blue) and without (black) the Lorentz force. The dotted line marks the collisional skin depth.

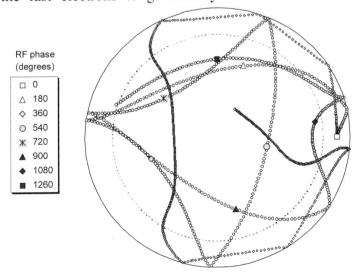

skimming the surface in the skin layer. Only when $\mathbf{F}_L$ is included do the electrons reach the interior with enough energy to ionize. This effect, which requires non-Maxwellian electrons as well as nonlinear forces, can explain why the density peaks at the center even when the skin layer is thin. Because of such non-classical effects, ICPs can be made to produce uniform plasmas even if they do not have antenna elements near the axis.

4. Ionization energy.

How much RF power does it take to maintain a given plasma density? There are three factors to consider. First, not all the power delivered from the RF amplifier is deposited into the plasma. Some is lost in the matchbox, transmission line, and the antenna itself, heating up these elements. A little may be radiated away as radio waves. The part deposited in the plasma is given by the integral of $\mathbf{J} \cdot \mathbf{E}$ over the plasma volume. If the plasma presents a large *load resistance* and the matching circuit does its job, keeping the reflected power low, more than 90% of the RF power can reach the plasma. Second, there is the loss rate of ion-electron pairs, which we have learned to calculate from diffusion theory. Each time an ion-electron pair recombines on the wall, their kinetic energies are lost. Third, energy is needed to make another pair, in steady state. The threshold energy for ionization is typically 15 eV (15.8 eV for argon). However, it takes much more than 15 eV, on the average, to make one ionization because of inelastic collisions. Most of the time, the fast electrons in the tail of the Maxwellian make excitation collisions with the atoms, exciting them to an upper state so that they emit radiation in spectral lines. Only once in a while will a collision result in an ionization. By summing up all the possible transitions and their probabilities, one finds that it takes more like 50–200 eV to make an ion-electron pair, the excess over 15 eV being lost in radiation. The graph of Fig. 22 by V. Vahedi shows the number of eV spent for each ionization as a function of KT_e.

5. Transformer Coupled Plasmas (TCPs)

As shown in Fig. 23, a TCP is an ICP with an antenna is shaped into a flat spiral like the heater on an electric stove top. It sits on top of a large quartz plate, which is vacuum sealed to the plasma chamber below containing the chuck and wafer. The processing chamber can also have dipole surface magnetic fields, and the antenna may also have a Faraday shield consisting of a

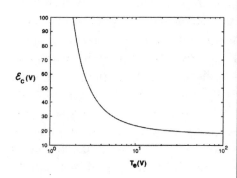

Fig. 22. The E_c curve, a very useful one in discussing energy balance and discharge equilibrium (L & L, p. 81).

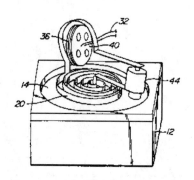

Fig. 23. Drawing of a TCP from the U.S. Patent Office.

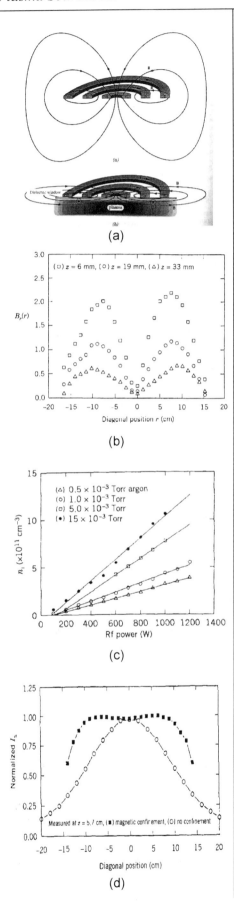

(a)

(b)

(c)

(d)

plate with radial slots. According to our previous discussion, the induced electric field in the plasma is in the azimuthal direction, following the antenna current. Electrons are therefore driven in the azimuthal direction to produce the ionization. As in other ICPs, the skin depth is of the order of a few centimeters, so the plasma is generated in a layer just below the quartz plate. Fig. 24a shows the compression of the RF field by the plasma's shielding currents. The plasma then diffuses downwards toward the wafer. In the vicinity of the antenna, its structure is reflected in irregularities of plasma density, but these are smoothed out as the plasma diffuses. There is a tradeoff between large antenna-wafer spacing, which gives better uniformity, and small spacing, which gives higher density. Being one of the first commercially successful ICPs, TCPs have been studied extensively; results of modeling were shown in Fig. 14. Densities of order 10^{12} cm^{-3} can be obtained (Fig. 24c). Magnetic buckets have been found to improve the plasma uniformity. There is dispute about the need for a Faraday shield: though a shield in principle reduces asymmetry due to capacitive coupling, it makes it harder to ignite the discharge. If the antenna is too long or the frequency too high, standing waves may be set up in the antenna, causing an uneven distribution of RF power. The ionization region can be extended further from the antenna by launching a wave—either an ion acoustic wave or an $m = 0$ helicon wave, but such TCPs have not been commercialized. The spiral coil allows the TCP design to be expanded to cover large substrates; in fact, very large TCPs, perhaps using several coils, have been produced for etching flat-panel displays.

TCPs and ICPs have several advantages over RIE reactors. There is no large RF potential in the plasma, so the wafer bias is not constrained to be high. This bias can be set to an arbitrary value with a separate oscillator, so the ion energy is well controlled. The ion energies also are not subject to violent changes during the RF cycle. These devices have higher ionization efficiency, so high ion fluxes can be obtained at low pressures. It is easier to cover a large wafer uniformly. In remote-source or detached-source operation, it is desired to have as little plasma in contact with the wafer as possible; the plasma is used only to produce the necessary chemical radicals. This is not possible with RIE devices. Compared with ECR machines, ICPs are much simpler and cheaper, because they require no magnetic field or microwave power systems.

Fig. 24. (a) RF field pattern without and with plasma; (b) radial profiles of B_r at various distances below the antenna, showing the symmetry; (c) density vs. RF power at various pressures; (d) plasma uniformity with and without a magnetic bucket. [from L & L, p. 400 *ff.*]

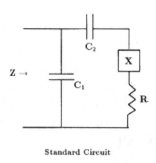

Standard Circuit

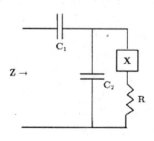

Fig. 25. The "standard" (left) and "alternate" (right) matching networks.

6. Matching circuits

The matching circuit is an important part of an ICP. It consists of two tunable vacuum capacitors, which tend to be large and expensive, mounted in a box with input and output connectors. At RF frequencies, wires are not simply wires, since every length of wire has an appreciable inductance; hence, the way the connections and ground plane are arranged inside the matchbox requires RF expertise. In addition, industrial plasma reactors have *automatch* circuits, which sense the way the capacitors have to be tuned to match the load and have little motors that automatically turn the tuning knobs. Design of matching circuits can be done with commercial network analyzers which plot out the *Smith chart* that is familiar to electrical engineers. However, we have derived analytic formulas which students can use without the expensive equipment (Chen, UCLA Report PPG-1401, 1992).

The capacitors, called the loading capacitor C_L (C_1 in the diagram) and the tuning capacitor C_T (C_2), can be arranged in either the standard or an alternate configuration. Let R_0 be the characteristic impedance of the power amplifier and the transmission lines (usually 50Ω), and let R and X be respectively the resistance and reactance of the load, both normalized to R_0. For instance for an inductive load with inductance L, X is $\omega L/R_0$. These will change when the plasma is created. For the standard circuit, the capacitances are

$$C'_L = [1-(1-2R)^2]/2R$$
$$C'_T = [X-(1-R)/C'_L]^{-1}, \qquad (6)$$

where $C_L' \equiv \omega C_L R_0$, etc. For the alternate circuit we have

$$C'_L = R/B, \qquad C'_T = (X-B)/T^2, \qquad (7)$$

where

$$T^2 \equiv R^2 + X^2, \qquad B^2 \equiv R(T^2 - R). \qquad (7a)$$

Note that these two circuits are actually identical: to switch from "standard" to "alternate", you merely have to swap the input and output terminals. The alternate circuits usually requires smaller capacitors with higher voltage ratings. These formulas do not include the cable that connects the matchbox to the antenna. Any cable length longer than a foot or so can make a big difference in the tuning; for instance, it can change in inductive load to a resistive load. This is because a ¼ wavelength of a 13.56 MHz wave in the cable is only about 5 m, and the reflected signal arriving back at the tuning circuit would be changed in phase significantly by a cable length of a fraction of a meter. How to handle transmission lines is covered in PPG-1401. Note in Eq. (6) that a capacitive load would have negative X, and then C_T would become negative; that is, it would require an inductor rather than a capacitor. Indeed, RIE discharges cannot be matched by purely capacitive tuning circuits; an inductor has to be added inside the matchbox. This inductor can be just a wire loop a couple of inches in diameter, but the coil must not be distorted or moved by the user, because that would change its inductance.

7. Electrostatic chucks (ESCs)

In plasma processing, photons and ions impinge on the wafer and heat it up. It is important to have an efficient way to remove the heat. A flow of He gas, a good heat transfer agent, is used on the back side of the wafer to cool the wafer. Originally, the wafer was held onto the chuck with mechanical fingers at the edges. This not only made the available area smaller but also allowed the wafer to bulge upwards under the He pressure, thus compromising its planarity. Electrostatic chucks have been developed to overcome these problems. In an ESC, a DC voltage is put on the chuck, charging it up. An opposite charge is attracted to the back side of the wafer, and these opposite charges attract each other, holding the wafer flat against the chuck at all points. To introduce the He, grooves are milled into the chuck. The back side of the wafers are rough enough that gas entering these grooves can seep beneath the wafer and remove heat from the whole surface.

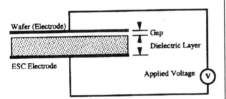

Fig. 26. (a) A monopolar ESC; (b) a bipolar ESC. What is seen is actually the coolant paths (from Applied Materials).

There are two types of ESC. Monopolar chucks have the same voltage applied to the entire chuck. The return path is through the plasma; that is, if a positive charge, say, is put on the chuck, the originally neutral wafer would have negative charges attracted to its back side and positive charges repelled to its front side. There, the charges are neutralized by electrons from the plasma. Thus, a monopolar chuck can function only if the plasma is on. There are problems with timing, because one has to make sure the wafer is not released too soon after plasma turnoff, and the wafer has to be firmly held as the plasma is turned on. The advantage of monopolar chucks is that it is easy to release the wafer. Bipolar chucks are divided into regions which get charged to opposite potentials. The return path in the wafer is

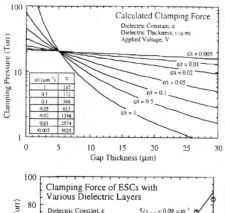

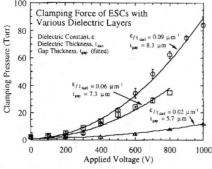

Fig. 27. Design curves for electrostatic chucks (from PMT, Inc.)

then through the wafer's cross section from one region to
the next. The plasma is not necessary for the chuck to
hold. The problem is that it is hard to get rid of the
charging after the process is over to release the wafer;
materials of just the right dielectric constant and conductivity have to be used in bipolar chucks.

An ESC consists of a flat plate (with grooves) covered with a layer of dielectric. On top of that is an air
gap, which is just the roughness of the wafer, and then
the wafer itself. The holding pressure of the chuck depends on the thickness t_{gap} of the air gap, and also on the
ratio ε / t, where ε is the dielectric constant of the dielectric layer of thickness t. For given values of these pa-
rameters, the holding pressure (in Torr) will increase
with chuck voltage. Dielectric constants vary from 3-4
for polyimide insulators to 9-11 for alumina or sapphire.
Chuck voltages can vary from 100 to 5000V. Since the
plasma potential oscillates at the RF and bias frequencies, the required chuck voltage can also vary with RF
power. Figure 27 shows the variation of clamping force
with gap thickness and voltage with ε / t.

PRINCIPLES OF PLASMA PROCESSING
Course Notes: Prof. F.F. Chen

PART A6: PLASMA SOURCES IV

VIII. HELICON WAVE SOURCES and HDPs

The newest type of High Density Plasma (HDP) is produced by the helicon wave source (HWS). This source requires a magnetic field of 50–1000G and is excited by an RF antenna, as in an ICP. The magnetic field has three functions: a) it increases the skin depth, so that the inductive field penetrates into the entire plasma; b) it helps to confine the electrons for a longer time; and c) it gives the operator extra adjustments to vary the plasma parameters, such as the density uniformity. The antenna launches a wave, called a helicon wave, that propagates along **B** with a phase velocity comparable to that of a 50-200 eV electron. The wave causes very efficient ionization, so that experiments are often done with densities in the mid-10^{13} cm^{-3} range, though such high densities are not usable for semiconductor processing. Nonetheless, HWS densities tend to be an order of magnitude higher than the 10^{11} cm^{-3} densities typical of ICPs. Until recently, it was not known why HW sources are so efficient. Initially, it was thought that cool electrons could catch the wave and surf on it up to the wave velocity, thus speeding up to where their ionization cross section was at its peak. Recent theories explain the efficient absorption of RF power by mode-coupling to another wave, a *Trivelpiece-Gould (TG) wave*, which will be described later.

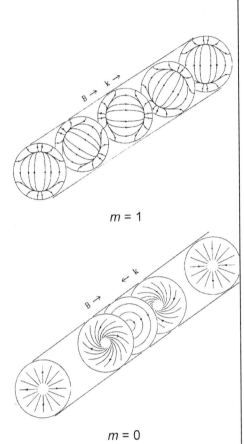

m = 1

m = 0

Fig. 1. Instantaneous E-field patterns for *m* = 1 and *m* = 0 helicon waves.

1. Dispersion relation

In Part A5, Eq. (A5-2) gave the dispersion relation for RH polarized e.m. waves propagating in the *z* direction (along **B**$_0$). If, instead, the wave vector *k* were at an angle θ to **B**$_0$, the dispersion relation would be

$$\frac{c^2 k^2}{\omega^2} = 1 - \frac{\omega_p^2}{\omega^2} \frac{1}{1 - \dfrac{\omega_c}{\omega} \cos\theta} \qquad (1)$$

This is called a *whistler wave* in space physics and was first heard by radio operators listening to ionospheric noise through headphones. At helicon densities and magnetic fields, the "1"s are both negligible, and the equation becomes

$$\frac{c^2 k^2}{\omega^2} = \frac{\omega_p^2}{\omega^2} \frac{\omega}{\omega_c \cos\theta} . \qquad (2)$$

Since $k^2 = k_z^2 + k_\perp^2$, the factor $\cos\theta$ is just k_z / k. Eq. (2) then becomes

$$k = \frac{\omega}{k_z} \frac{\omega_p^2}{\omega_c c^2} = \frac{\omega}{k_z} \frac{en\mu_0}{B} . \qquad (3)$$

This shows how the basic helicon dispersion relation is related to the ones we have already encountered. To be consistent with helicon terminology, however, we now make a few changes in notation. The k above is the total propagation constant, and it will be called β from now on. The z component k_z will now be called k. Eq. (3) then becomes

$$\beta = \frac{\omega}{k} \frac{ne\mu_0}{B}, \qquad \beta^2 \equiv k_\perp^2 + k^2 . \qquad (4)$$

Since helicon waves are whistlers confined to a cylinder, the plane-geometry concept of $k_\perp$ is no longer useful; $k_\perp$ will later be replaced by arguments of Bessel functions. In cylindrical geometry, $k_\perp$ is approximately $3.83/a$, where a is the plasma radius, the 3.83 coming from a Bessel function root. In basic experiments, the radius of the helicon plasma is much smaller than its length, and one has $k_\perp \gg k$ and $\beta \approx k_\perp$. We see from Eq. (4) that if one fixes ω, the radius a, and the wavelength $2\pi/k$ (by adjusting the length of the antenna), then n/B must be fixed. In the simplest helicon wave, *the density should increase linearly with magnetic field.*

2. Wave patterns and antennas

If we make the simplifying assumption that the plasma is uniform in the z and θ directions, we can Fourier analyze and study each z and θ mode by assuming that each wave quantity, such as **B** (as distinguished from the dc field $B_0\hat{\mathbf{z}}$), varies as

$$\mathbf{B} = \mathbf{B}(r)e^{i(m\theta + kz - \omega t)} . \qquad (5)$$

Thus, the wave propagates in the z direction with a wavelength $2\pi/k$, and its amplitude varies in the θ direction as $\cos(m\theta)$. Here m is the *azimuthal wave number*. Thus, $m = 0$ is a mode that is azimuthally symmetric,

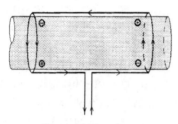

Nagoya Type III

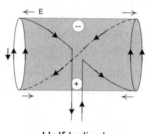

Half helical

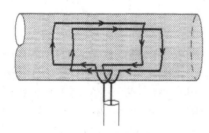

Double saddle coil

Fig. 2. Common types of helicon antennas.

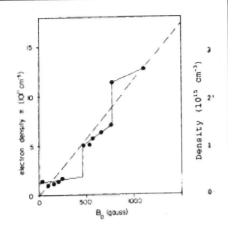

Fig. 3a. Density jumps with increasing field (from Boswell et al.).

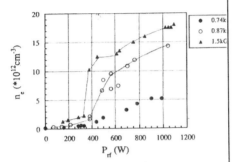

Fig. 3b. Density jumps with increasing power (from Shoji et al.).

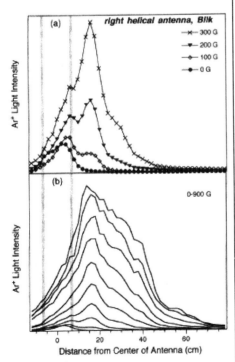

Fig. 4. Increase of density with B-field.

while $m = 1$ is a mode that varies as $\cos\theta$ and is RH polarized (θ increases as t increases). Similarly, $m = -1$ is LH polarized. Though LH waves in free space are evanescent, both RH and LH waves can propagate in a cylinder. Actually LH waves are not easily excited, for obscure reasons.

The electric field patterns for $m = 1$ and $m = 0$ helicons are shown in Fig. 1. The $m = 1$ mode has a pattern that does not change as it rotates. As the wave propagates in the z direction (the direction of **B**), a stationary observer would see the pattern rotating clockwise as viewed along **B**. The $m = 0$ mode is entirely different: the pattern is not invariant but changes from electrostatic (radial E-lines) to electromagnetic (azimuthal E-lines) in each half-cycle.

Antennas can be shaped to launch particular modes. Some are shown in Fig. 2. For instance, a simple hoop or two separated hoops with current in opposite directions can launch $m = 0$ modes. A Nagoya Type III (N3) antenna has two parallel legs connecting two such hoops; these legs are actually more important than the end rings. The N3 launches plane-polarized $|m| = 1$ modes consisting of equal amplitudes of $m = +1$ and $m = -1$. In practice, however, the $m = -1$ mode can hardly be detected, and the $m = +1$ mode is the one that is launched in both directions. A half-helical (HH) antenna is an N3 antenna with one end twisted 180° to match the helicity of the $m = 1$ wave pattern. Like the N3, it is half a wavelength long, chosen to give the right value of k_z in Eq. (3). Depending on the direction of **B**$_0$, the HH launches an $m = +1$ wave out one end and a (small) $m = -1$ wave out the other. The double saddle coil is an N3 antenna with each parallel leg spit into two paths, slightly separated. The advantage of this is that it can be slipped onto a tube without breaking vacuum. With helical antennas, one does not really have to build antennas with different directions of twist. To change between $m = +1$ and $m = -1$ excitation, one merely needs to reverse the magnetic field. This can be seen from Eq. (5), since the z direction is defined to be that of **B**$_0$.

3. Mode jumping

At low **B**$_0$ or low power P_{rf}, the conditions are not right for generating helicon waves, and one gets only a low density plasma characteristic of ICPs. As the field is raised above 200-400G (depending on other parameters such as pressure), the helicon dispersion can be satisfied, and the density discontinuously jumps to a high value

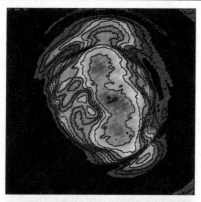

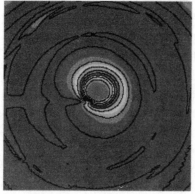

Fig. 5. Axial view of a helicon discharge in 488 nm Ar$^+$ light before and after jump into the big blue mode (which is color-coded red here!).

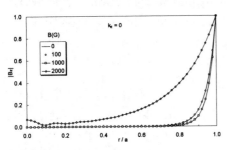

Fig. 6a. Decay of a $k_z = 0$ antenna field vs. $\mathbf{B}_0$.

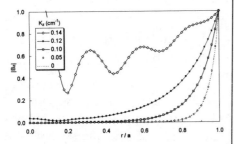

Fig. 6b. Decay of antenna field at $\mathbf{B}_0$ = 100G for various k_z.

satisfying Eq. (4) (Fig. 3a). From then on, n increases linearly with B. Similar jumps are seen as P_{rf} is increased (Fig. 3b). At low power, there is only capacitive coupling, and the density is very low. As the power is raised to 100-300 W, inductive coupling takes hold, and the density jumps to a value characteristic of ICPs. At 400 W or so, the helicon mode is struck, and the density takes another jump to a value satisfying (4). In Fig. 4, we see that, as $\mathbf{B}_0$ is raised from 0, the density on axis can increase by 20 times over its $\mathbf{B}_0 = 0$, or ICP, value. However, the averaged density is not changed quite as much. What happens is that the plasma snaps into a "big blue" mode with a bright, dense central core. This core has a higher T_e and higher ionization fraction than the surrounding plasma. For processing purposes, however, the $>10^{13}$ cm^{-3} densities of the blue mode at 1000G and 2kW are not useful. Lower $\mathbf{B}_0$ and P_{rf} are used to create a weaker, pinkish (in argon) plasma which will dissociate a molecule like Cl_2 but not totally ionize it.

4. Modified skin depth

The skin depth in ICPs is determined by shielding current of electrons. With as little as 10G of magnetic field, the electrons would have small Larmor radii and be unable to flow in the direction required for shielding. One would think that the rf field would then penetrate easily into the plasma. However, because of a mechanism too complicated to explain here, this does not happen. If $k_z = 0$, the skin depth does not change appreciably until $\mathbf{B}_0$ reaches 1000s of gauss. If the antenna field is not constant over all z — that is, if $k_z \neq 0$ — the skin depth can again be increased by the B-field. The rf field, however, still does not penetrate all the way to the axis until k_z is large enough to satisfy Eq. (3) so that helicon waves can be excited. Thus, helicon waves are necessary for getting rf energy to the center of the plasma. The anomalous skin depth mechanism of ICPs will not work in a magnetic field.

5. Trivelpiece-Gould modes

With the same amount of power deposited into the plasma, helicon discharges produce more ions than do ICP discharges; it is not just the peak density that is higher. This was hard to understand, since the usual transfer of rf energy to the plasma is through collisional damping of the wave fields, and it can be shown that this damping is quite weak for helicon waves. Magnetic confinement helps a little (not nearly enough); on the other hand the field prevents the drift of fast electrons into the center, as in ICPs. At first, it was thought that *Landau damping* was responsible for the efficient energy transfer. This is a mechanism in which electrons surf on the wave and gain energy up to the phase velocity.

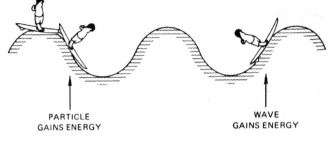

For usual values of n and B in Eq. (4), the phase velocity ω / k would put the surfing electrons at an energy near the maximum of the ionization cross section (~100 eV), thus increasing the ionization rate. This was such an attractive explanation that numerous experiments were performed to prove it. Careful measurements of the EEDF, however, showed no such fast-electron tail. This collisionless damping mechanism could still occur near the antenna, but the current belief is that it is insufficient to account for the high density of helicon discharges.

An alternative explanation was found when the mass ratio m / M, though small, was assumed to be finite instead of zero, as it was in deriving Eq. (3). Two waves are then found which satisfy the differential equations

$$\nabla^2 \mathbf{B} + \beta_1 \mathbf{B} = 0, \quad \nabla^2 \mathbf{B} + \beta_2 \mathbf{B} = 0, \qquad (6)$$

where β_1 and β_2 are the total wave numbers of the two waves. The β's satisfy a quadratic equation, whose roots are

$$\beta_{1,2} = \frac{k}{2\delta}\left[1 \mp \left(1 - \frac{4\beta_0 \delta}{k}\right)^{1/2}\right], \qquad \delta \equiv \frac{\omega}{\omega_c}. \quad (7)$$

Here β_0 is the approximate value of β given by Eq. (4), and δ is not the skin depth δ !. The upper sign gives $\beta_1 \approx \beta_0$, the modified helicon wave, which approaches β_0 as $\delta \to 0$. The lower sign gives $\beta_2 \approx k/\delta$, the *Trivelpiece-Gould* (TG) mode, which is essentially an electron cyclotron wave confined to a cylinder. For given k, both waves exist at the same time, and their β values are shown in Fig. 7 for various values of $\mathbf{B}_0$.

In Fig. 7, for a given value of k, there are two possible β's: β_1 and β_2. The smaller one has longer radial

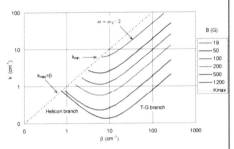

Fig. 7. The $k - \beta$ curves for combined helicon–TG modes.

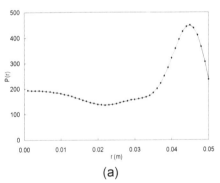

(a)

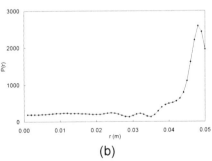

(b)

Fig. 8. Radial profiles of energy deposition at (a) 50G and (b) 300G showing the relative contributions of the H and TG modes.

Fig. 9. The glow of a helicon discharge [U. of Wisconsin].

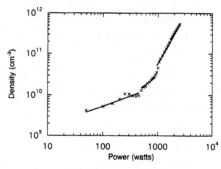

Fig. 10. Transition between capacitive coupling (E-mode), ordinary inductive coupling (H-mode), and helicon coupling (W-mode) [Degeling et al., Phys. Plasmas **3**, 2788 (1996)].

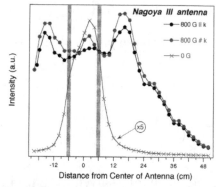

Fig. 12. A Nagoya III antenna launches $m = +1$ waves in both directions, as long as $\mathbf{B}_0$ is above the threshold for the W-mode.

wavelength and is the helicon (H) wave; the larger β has shorter radial wavelength and is the TG wave. Its wavelength can be so short that it damps out before going far into the plasma, and it is then essentially a surface wave. Nonetheless, because it is damped rapidly, it accounts for the efficient RF power absorption of the HWS. The mechanism is as follows. The antenna excites the H wave as usual, but the H wave is damped so weakly that it cannot account for all the RF power absorbed. For $\delta > 0$, however, the H wave alone cannot satisfy the boundary condition at $r = a$; a TG wave must be generated there as well, with an amplitude determined by the boundary condition. The TG wave is heavily damped and deposits RF energy near the surface. At low B-fields, the curve becomes shallower, and β_1 and β_2 are close to each other, so that the H and TG waves are strongly coupled, and both extend far into the plasma. For $\omega_c < 2\omega$, the H wave becomes evanescent, and only the TG wave can propagate. At high $\mathbf{B}_0$, note that there is a minimum value of k; that is, the axial wavelength cannot be overly long.

6. Examples of helicon measurements

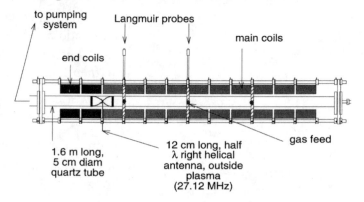

Fig. 11. A uniform-field system used for most of the studies of helicon discharges shown here [UCLA].

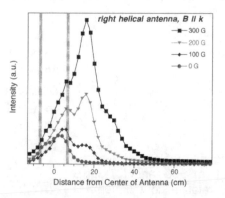

Fig. 13. A helical antenna launches the $m = +1$ mode in one direction only.

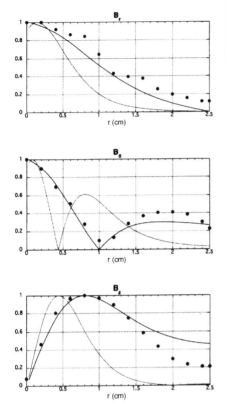

Fig. 15. Measured radial density profiles of B_i, compared with the Bessel solutions for $m = +1$ and $m = -1$ (faint line). Regardless of the direction of B_0, it was not possible to get profiles agreeing with $m = -1$.

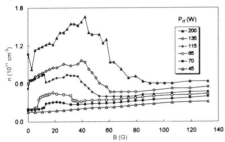

Fig. 16. The low-field density peak for various P_{rf}.

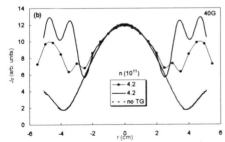

Fig. 17. Radial profile of the RF current (points) at 40G, as compared with simple helicon theory (bottom curve) and the theory including the TG mode.

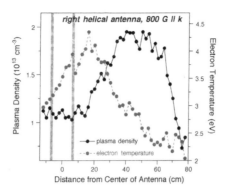

Fig. 14. With a helical antenna, both n and KT_e peak downstream of the antenna (at the left, between the bars).

The data shown above illustrate the behavior of helicon discharges in "normal" operation, at fields high enough that Eq. (3) is obeyed, and n increases linearly with B_0. Peak densities of up to 10^{14} cm^{-3} have been observed. Though the TG mode is not seen because it is localized to a thin boundary layer at these fields, it is nonetheless responsible for most of the energy absorption. Helicons can also be generated at fields of ~100G and below, exhibiting behavior different from ICPs. A low-field density peak is often seen, with the density decaying before it goes up linearly with B_0 in its high-field behavior (Fig. 16). The densities in this peak are not high but are still higher than in ICPs. It is believed that this peak is due to constructive interference from the wave reflected from a nearby endplate. At low B-fields, the TG mode has a comparatively long radial wavelength and can be seen in the RF current J_z, which emphasizes the TG mode. The TG mode generates peaks in J_z near the boundary, as seen in Fig. 17. These additional peaks are not predicted by the theory neglecting the TG modes.

7. Commercial helicon sources

Two commercial helicon reactors have been marketed so far: the Boswell source, and the MØRI (M = 0 Reactive Ion etcher) source of PMT, Inc. (now Trikon). Variations of the Boswell source have been sold by various companies on three continents, including Lucas Labs in the U.S. The Lucas source uses a paddle (or saddle)-shaped $m = 1$ antenna. The MØRI source uses a two-loop $m = 0$ antenna with the currents in opposite directions. It incorporates a magnetic bucket in the processing chamber to confine the primary electrons. The field coils have a special shape: two coils in the same plane with currents in opposite directions. These opposed coils

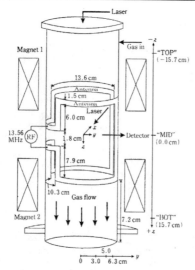

Fig. 18. The Lucas source.

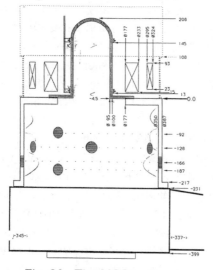

Fig. 20. The MØRI source.

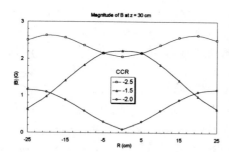

Fig. 22. Computed B-field strength at wafer level for three values of the coil current ratio.

make the magnetic field diverge rapidly, so that very little field exists at the wafer. The density profile can be flattened by controlling the ratio of the currents in the two coils. The magnetic fields used in these reactors is in the 100-400G range.

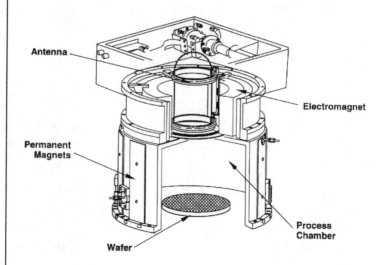

Fig. 19. Drawing of the PMT MØRI source.

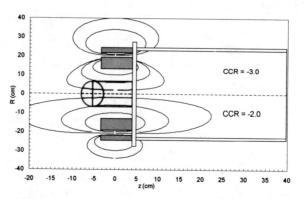

Fig. 21. The B-field patterns in the MØRI source (on its side) for various coil current ratios (CCR). A negative ratio bends the field lines away from the substrate and can be adjusted to give < 1G at the wafer level.

To cover large substrates, multiple helicon sources are being developed at UCLA. These comprise an array of appropriately spaced tubes, each being short to make use of the low-field peak mentioned above. Uniform densities above 10^{12} cm^{-3} with ±3% uniformity over a 40-cm diameter have been achieved. In this example, six tubes are spaced around a central tube. Density scans over the cross sectional area showed no six-fold asymmetry due to the individual sources.

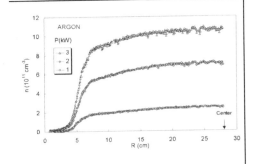

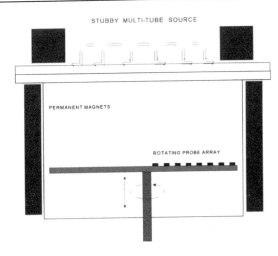

Fig. 23. A multi-tube *m* = 0 helicon source

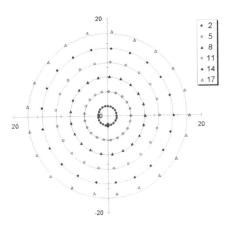

Fig. 25. Density variations at various radii.

Performance of the distributed source of Fig. 23 is shown in Figs. 24 and 25, and in Sec. A3, under Uniformity.

IX. DISCHARGE EQUILIBRIUM (L&L, p. 304*ff*)

In high-density plasmas (HDPs), the plasma density, neutral density, bulk electron temperature, and EEDF determine the performance of the reactor in producing the right mixture of chemical species and the energy and flux of ions onto the substrate. For some processes, such as cleaning and stripping, simply a high density of oxygen would do; but for the more delicate etching processes, the proper equilibrium conditions are crucial. Reactor-scale numerical modeling can treat in detail the exact geometry of the chamber and the gas feeds, and the radial profiles of the various neutral and ionized species. In more detailed computations, even the sheath and non-Maxwellian electron effects are included. The problem with modeling is that only one condition can be computed at a time, and it is hard to see the scaling laws behind the behavior of the discharge. To get an idea of the relation between energy deposition and the resulting densities and temperatures, we need to examine particle and energy balance in a plasma.

1. Particle balance

Consider the rate at which plasma is lost to the walls in an arbitrary chamber of volume V and surface area A. Since electrons are confined by sheaths, the loss rate is governed by the flux of ions through the sheath edge at the acoustic speed c_s, according to the Bohm criterion (A1-19). If N is the total number of ions, it will decrease at the rate

$$\left.\frac{dN}{dt}\right|_{out} = An_s c_s = A g_1 \tfrac{1}{2}\bar{n} c_s \approx \tfrac{1}{2} A \bar{n} c_s. \qquad (8)$$

Here g_1 is a geometric factor of order unity relating the sheath-edge density to the average density. It differs from ½ in realistic plasmas which are not entirely uniform in the quasineutral region. In steady state, N is replenished by ionization. The ionization rate is $n_n n <\sigma v>_{ion}$, so N is increased at the rate

$$\left.\frac{dN}{dt}\right|_{in} = V g_2 n_n \bar{n} <\sigma v>_{ion}, \qquad (9)$$

where g_2 is another geometric factor of order unity to account for density inhomogeneity. Equating input to output, $\bar{n}$ cancels; and we have

$$\frac{1}{2}\frac{g_1}{g_2}\frac{A}{V} = n_n \left[\frac{<\sigma v>_{ion}}{c_s}\right] \equiv n_n f(T_e), \qquad (10)$$

where $f(T_e)$ is the function of T_e in the square brackets. The left-hand side depends only on the geometry of the plasma. For instance, if the chamber is a cylinder of radius R and length L, the area is $2\pi R^2 + 2\pi RL$, while the volume is $\pi R^2 L$. The l.h.s. is then approximately $(R + L)/RL$, or simply $1/R$ for $L \gg R$. Thus, for a given geometry and ion mass, T_e is related to the neutral density n_n and is independent of plasma density n. This unique relationship is shown in Eq. 26 for $R = 16$ cm.

Thus, particle balance—that is, equating the rate of plasma production with the rate of loss—does not give a value of the density n. We have only a relation between KT_e and n_n. If n_n is depleted by strong ionization, the abscissa in this graph (the filling pressure p_0) should be replaced by the local pressure $p(\text{mTorr}) = n_n / 3 \times 10^{13}$ cm^{-3}. This depletion is likely to vary with position; for instance, near the axis, and then one would expect a local rise in KT_e there. Solving the diffusion equations is necessary only to give the geometrical factors in Eq.(10), which could be used to refine the $T_e - p$ relation.

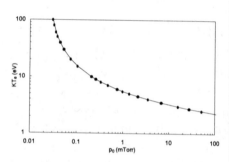

Fig. 26. A T_e vs. p_0 curve.

2. Energy balance

The equilibrium density of the plasma can be estimated from the absorbed RF energy. This is given by Ohm's law, essentially $I^2 R$, or actually $\mathbf{E \cdot J}$ integrated over the volume of the plasma and averaged over time. It is almost equal to $\mathbf{E \cdot J}$ integrated over the antenna, which can be measured by multiplying the voltage and current applied to the antenna times the cosine of the angle between them. The only difference is the power dissipated by the antenna's resistance and the power radiated away into space (both of these effects are negligible). This energy in must be balanced with the energy out.

The plasma loses energy by particle loss and by radiation of spectral lines. The total energy lost when each ion-electron pair recombines at the wall can be divided into three terms:

$$W_{tot} = W_i + W_e + W_c. \tag{11}$$

W_i is the energy carried out by the ion after falling through the sheath drop V_{sh}. Since it reaches the sheath edge with a velocity c_s, its energy there is $\frac{1}{2}KT_e$. Thus we have

$$W_i = \frac{1}{2}KT_e + eV_{sh}. \tag{12}$$

W_e is the energy carried out by the electron and on average is equal to $2KT_e$. The reason this is not simply KT_e or $(3/2)KT_e$ is that the electron is moving at a velocity v_x toward the wall, but it also carries energy due to its v_y and v_z motions. We can derive this as follows. The flux of electrons to the wall is

$$\Gamma_e = n \int_0^\infty v_x e^{-mv_x^2/KT_e} dv_x. \tag{13}$$

Each carries an energy

$$E(v_x) = \frac{1}{2}mv_x^2 + <\frac{1}{2}mv_\perp^2> = \frac{1}{2}mv_x^2 + KT_e. \tag{14}$$

The reason the average energy in the y and z directions is KT_e is that there is $\frac{1}{2}KT_e$ of energy in each of the two degrees of freedom. The flux of energy out, therefore is the same as Eq. (13) but with an extra factor of $E(v_x)$. The mean energy carried out by each electron is then

$$W_e = \frac{\int_0^\infty E(v_x)\exp(-mv_x^2/2KT_e)v_x dv_x}{\int_0^\infty \exp(-mv_x^2/2KT_e)v_x dv_x}. \tag{15}$$

Since the KT_e term of Eq. (14) does not depend on v_x, it factors out, and we have

$$W_e = KT_e + \frac{\int_0^\infty (\tfrac{1}{2}mv_x^2)\exp(-mv_x^2/2KT_e)v_x dv_x}{\int_0^\infty \exp(-mv_x^2/2KT_e)v_x dv_x} = KT_e + \frac{KT_e \int_0^\infty e^{-w}w\,dw}{\int_0^\infty e^{-w}dw} \qquad (16)$$

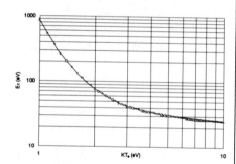

Fig. 27. The Vahedi curve shown in detail (data from P. Vitello, LLNL).

In the last step we have introduced the abbreviation

$$w \equiv mv_x^2/2KT_e, \qquad dw = (m/KT_e)v_x dv_x.$$

Integrating by parts, we have $\int U dV = UV - \int V dU$, where

$$U = w \qquad dU = dw$$
$$dV = e^{-w}dw \qquad V = -e^{-w}$$

The integral in the numerator becomes

$$\int_0^\infty e^{-w}w\,dw = -we^{-w}\Big|_0^\infty + \int_0^\infty e^{-w}dw.$$

The first term vanishes, and the second term is 1; in any case, it cancels the denominator. Thus, Eq. (16) gives

$$W_e = 2KT_e. \qquad (17)$$

The last term, W_c, in Eq. (11) is the same as the function $E_c(T_e)$ introduced in Sec. VII-4 of Part A5, is the energy required to produce each electron-ion pair, including the average energy radiated away in line radiation during this process. In steady state, every ion lost has to be replaced, and therefore this ionization and excitation energy is lost also. The E_c curve, originally due to Vahedi, is important, since it represents the major part of the energy loss. The useful part of this curve is given here on a log-log scale for argon.

Equating the power lost to the power P_{rf} absorbed by the plasma, we have, from Eq. (8),

$$P_{rf} = \tfrac{1}{2}ng_1 c_s A W_{tot}, \qquad (18)$$

from which the density can be calculated.

3. Electron temperature

In the previous discussion, we assumed a more or less uniform and steady-state plasma with ionization occurring throughout. The plasma, however, can be non-steady in several ways. It can be pulsed so as to make use of the afterglow, where KT_e drops to a low value. It can also flow away from the antenna to form a detached plasma. If there is no more energy input in the downstream region, the electron temperature is determined by a balance between energy loss by radiation and energy gain by heat conduction. We give an example of this. The heat conduction equation for the electron fluid is

$$\frac{d}{dt}\left(\frac{3}{2}nKT_e\right) = Q - \nabla \cdot \mathbf{q} = 0, \qquad (19)$$

where Q is the heat source, and **q** the heat flow vector, and the time derivative is zero in steady state. Because of their small mass, electrons lose little energy in colliding with ions or neutrals, so Q is essentially $-W_c$. For flow along $\mathbf{B}_0$, q is given in Coulomb scattering theory by

$$q = -\kappa_{\parallel}\frac{\partial}{\partial z}KT_e, \quad \kappa_{\parallel} = 3.2\frac{nKT_e}{m}\tau_e,$$

$$\tau_e = 3.44\times10^5\frac{T_{eV}^{3/2}}{n\ln\Lambda}, \qquad (20)$$

where $\kappa_{\parallel}$ is the coefficient of heat conduction along $\mathbf{B}_0$ and τ_e is the electron-electron scattering time. A solution of Eq. (19) can be compared with the KT_e data given above for a helicon discharge, and the result shows good agreement (Fig. 28).

4. Ion temperature

The ion temperature does not play an important role in the overall equilibrium, and it is difficult to measure. However, it is interesting because anomalously high KT_i has been observed in helicon discharges, and this has not yet been definitively explained. Here we wish to see what KT_i *should* be according to classical theory. The ions gain energy by colliding with electrons, which get their energy from the RF field. The ions lose energy by colliding with neutrals. Since the neutrals have almost the same mass as the ions, the latter process will dominate, and T_i is expected to be near T_0, the temperature of the neutrals. The rate of energy gain is proportional to the difference between T_e and T_i:

$$\frac{dT_i}{dt} = v_{eq}^{ie}(T_e - T_i), \qquad (21)$$

where the coefficient is the rate of equilibration between ions and electrons and is obtained from the theory of Coulomb collisions. It is proportional to $T_e^{-3/2}$. The rate

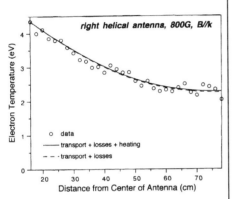

Fig. 28. Data and theory showing electron cooling by inelastic collisions (UCLA).

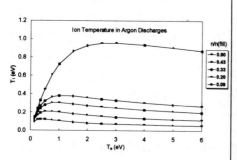

Fig. 29. Ion temperature vs. p_0 and KT_e, according to classical Coulomb theory.

of energy loss is proportional to the difference between T_i and T_0:

$$-\frac{dT_i}{dt} = n_n <\sigma v>_{cx} (T_i - T_0) , \qquad (22)$$

where the charge-exchange collision rate is used because that is dominant. Equating the loss and gain rates, we can solve for T_i as a function of T_e and p_0. One such solution is shown in Fig. 29. We see that for fractional ionizations <10%, as are normal, T_i should be < 0.1 eV. It is therefore surprising that KT_i's of order 1 eV have been observed. This could happen if there is neutral depletion, so that the neutral density is far below the fill density. It has also been suggested that the ions are heated by low-frequency waves generated by an instability.

PRINCIPLES OF PLASMA PROCESSING
Course Notes: Prof. F.F. Chen

PART A7: PLASMA DIAGNOSTICS

X. INTRODUCTION

Diagnostics and *sensors* are both measurement methods, but they have different connotations. Diagnostic equipment is used in the laboratory on research devices and therefore can be a large, expensive, and one-of-a-kind type of instrument. Sensors, on the other hand, are used in production and therefore have to be simple, small, unobtrusive, and foolproof. For instance, *endpoint detectors*, which signal the end of an etching step by detecting a spectral line characteristic of the underlying layer, are so important that they are continually being improved. Practical sensors are few in number but constitute a large subject which we cannot cover here. We limit the discussion to laboratory equipment used to measure plasma properties in processing tools.

Diagnostics for determining such quantities as n, KT_e, V_s, etc. that we have taken for granted so far can be remote or local. Remote methods do not require insertion of an object into the plasma, but they do require at least one window for access. Local diagnostics measure the plasma properties at one point in the plasma by insertion of a probe of one type or another there. Remote methods depend on some sort of radiation, so the window has to be made of a material that is transparent to the wavelength being used. Sometimes quartz or sapphire windows are needed. The plasma can put a coating on the window after a while and change the transmission through it. Probes, on the other hand, have to withstand bombardment by the plasma particles and the resulting coating or heating; yet, they have to be small enough so as not to change the properties being measured.

XI. REMOTE DIAGNOSTICS

1. Optical spectroscopy

One common remote diagnostic is optical emission spectroscopy (OES), which is the optical part of the more general treatment of radiation covered in Part B. In OES, visible light is usually collected by a lens and focused onto the slit of a spectrometer. The detector can be a photodiode, a photomultiplier, or an optical multichannel analyzer (OMA). With a photodiode, interference filters are used to isolate a particular spectral line. Optical radiation can also be used to image a plasma in the light of a particular spectral line using an interference

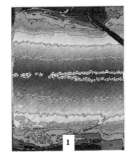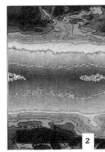

Fig. 1. Emission of ionized argon light at various *z* positions in a helicon discharge.

filter and a sensitive CCD camera. Fig. 1 shows the emission from ionized argon recorded with a narrow-band filter for the 488 nm line of Ar^+.

A photomultiplier can see only one part of the spectrum at a time, but it is the most sensitive detector for faint signals. An OMA records an entire range of wavelengths on a CCD (charge-coupled detector) and is the convenient for scans of a single line or for recording an entire spectrum.

By comparing the intensities of different spectral lines, one can determine not only the atomic species present but also the electron temperature, density, and the ionization fraction. The relative intensities of two lines with different excitation thresholds can yield KT_e. The relative intensities of an ion line and a neutral line can be used to estimate the ionization fraction. In principle, line broadening contains a large amount of information, but only for hot, highly ionized plasmas. For instance, *Doppler broadening* yields the velocity of the emitting ion or atom. *Stark broadening* or *pressure broadening* gives information on density. This is because, at high densities, collisions interrupt the emission of radiation, and hence the line cannot contain a single frequency. In plasma processing, the most useful and well developed technique is *actinometry*. In this method, a known concentration of an impurity is introduced, and the intensities of two neighboring spectral lines, one from the known gas and one from the sample, are compared. Since both species are bombarded by the same electron distribution and the concentration of the actinometer is know, the density of the sample can be calculated.

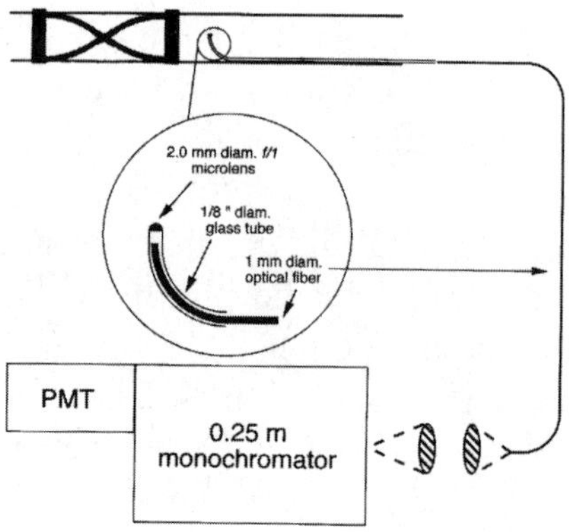

Though most optical methods average over a ray path in the plasma, a more local measurement of light emission can be made with a probe containing a small lens coupled to an optical fiber. Such a probe is shown in Fig. 2, and data from it in Fig. 3.. The lens collects light preferentially from a small focal spot just in front of it. The Ar^+ light collected by it is localized under the antenna if $\mathbf{B}_0 = 0$, as would be expected in ICP operation.

2. Microwave interferometry

Another useful remote diagnostic is *microwave interferometry*. A beam of microwave radiation is launched by a horn antenna into a plasma through a window. According to Eq. (A5-1), these waves can propagate in the plasma if $\omega > \omega_p$. From (A5-1) it is easily seen that the phase velocity in the plasma is

Fig. 2. Schematic of a local OES probe.

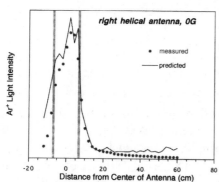

Fig. 3 Example of data on optical emission vs. z.

$$\frac{\omega}{k} = \frac{c}{(1 - \omega_p^2/\omega^2)^{1/2}}. \tag{1}$$

This is faster than the velocity of light, but it is quite all right for *phase* velocity to be $> c$ as long as the group velocity is $< c$. The microwave beam therefore has a longer wavelength inside the plasma than in air. The presence of the plasma therefore changes the phase of the microwave signal, a change which increases with the density of the plasma. The "standard" setup is shown in Fig. 4. The microwave beam from a generator is split into two parts, one going through the plasma and the other going through a waveguide toward the detector, where the two beams are recombined By adjusting the reference signal with an attenuator and phase shifter, the two signals can be made to cancel each other, so that the detector shows zero signal when there is no plasma. If the plasma density is increased slowly, the signal going through the plasma will have undergone fewer oscillations, and this phase shift will cause the nulled detector to give a finite dc signal output. If the plasma density reaches a value such that the wave loses exactly one wavelength, the detector will again return to zero; the signal is shifted by one *fringe*. By counting the number of fringes either on the way up to maximum density or on the way down, one obtains a measure of the average density traversed by the microwave beam. Though this illustrates the principle of interferometry, it is not normally done this way. First, ω is usually chosen so that ω_p/ω is a small number; then, the phase shift is linearly proportional to n. Second, the entire reference leg can be replaced by a mirror on the opposite side of the plasma to reflect the beam back into the launching horn. The beam then travels twice through the plasma and suffers twice the phase shift. Besides increasing the sensitivity, this method obviates phase shifts in the reference leg due to small changes in room temperature, which change the length of the waveguide. If the plasma is always on, it is difficult to set the initial null of the detector. There are various ways to get around this which we need not explain here. Modern network analyzers can do most of these calibrations automatically, but the principle of operation is always the same.

The phase shift $\Delta\phi$ that the plasma causes can be calculated as follows. If $k_0 = \omega/c$ is the propagation constant in air and k_1 is that in the plasma, we have

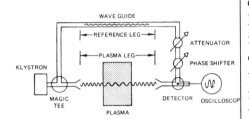

Fig. 4. Schematic of a microwave interferometer (Chen, p. 91).

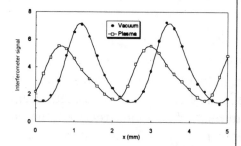

Fig. 5. Fringe shifts as the path length is changed. The lines are analytic fits through the points.

Fig. 6. Fringe patterns views along the axis can show the shape of the plasma [Heald and Wharton, 1978].

$$\Delta\phi = \int (k_0 - k_1)dx , \qquad (2)$$

where k_1 is given by Eq. (1) as

$$k_1 = k_0\left(1 - \frac{n}{n_c}\right)^{1/2} . \qquad (3)$$

Here we have replaced ω_p^2/ω^2 by n/n_c, where n_c is the "critical density" defined by

$$\omega^2 = \frac{n_c e^2}{\varepsilon_0 m} . \qquad (4)$$

The phase shift is then

$$\Delta\phi = k_0 \int \left[1 - \left(1 - \frac{n(x)}{n_c}\right)^{1/2}\right] dx . \qquad (5)$$

We see that the phase shift measures only the *line integral* of the density, not the local density. If ω is high enough that $n << n_c$, Eq. (5) can be Taylor expanded to obtain

$$\Delta\phi \approx \frac{k_0}{2n_c} \int n(x)dx \approx \tfrac{1}{2} k_0 L \frac{<n>}{n_c} \text{ radians}, \qquad (6)$$

where $<n>$ is the average density over the path length L. In the reflection method, the integral (or L) must be doubled. Fig. 5 gives an example of the interferometer output in the double-pass method as the mirror is moved to change the path length. The fringe shift is clearly seen, but it is also evident that the waveform has been distorted. This is because the microwave generator did not give a pure signal, and its harmonics at higher ω suffered a different phase shift. By fitting the curves to sine waves and their harmonics and adjusting the relative phases, one can recover the phase shift of the fundamental and thus get the density. Fig. 6 shows an end view of a dense plasma, in which the path length was so long that many fringes are seen, revealing the shape of the plasma.

Microwave interferometry is useful for calibrating Langmuir probes. With a probe, one can measure the density profile across a radius or diameter of the plasma, but the absolute value of the density may not be known accurately. By using the measured density profile to compute the integral in Eq. (6), one can find the absolute density by measuring the microwave phase shift. The errors in this method come from the fact that the plasma is not perfectly planar, and the microwave beam is not

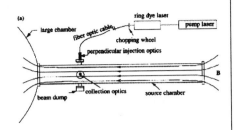

Fig. 7. Perpendicular alignment of injection laser and collection optics [Scime et al., Plasma Sources Sci. Technol. **7**, 186 (1998)].

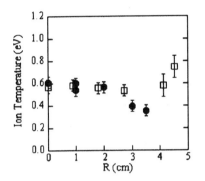

Fig. 8. LIF data on T_i parallel (solid points) and perpendicular (open points) to $\mathbf{B_0}$, showing anomalously high KT_i [Kline et al., Phys. Rev. Lett. **88**, 195002 (2002)].

perfectly parallel. Refraction can cause part of the beam to miss the collector, and reflections from the chamber walls can cause spurious waves.

3. Laser Induced Fluorescence (LIF)

This diagnostic is both non-invasive and local because it uses intersecting beam paths. Furthermore, it is the only way to measure T_i without using a large energy analyzer. One laser, tuned to a particular transition, is used to raise ions to an excited state along one path through the plasma. The excited ions fluoresce, giving off light at another frequency, and this light is collected by a lens focused to one part of the path, providing the localization. Doppler broadening of the line yields the ion velocity spread in a particular direction. The equipment is large, expensive, and difficult to set up, so that it is available in a relatively few laboratories. LIF is treated in more detail in Part B. Figure 7 shows a typical LIF setup, and Fig. 8 shows data taken in a helicon plasma.

XII. LANGMUIR PROBES

1. Construction and circuit

A Langmuir probe is small conductor that can be introduced into a plasma to collect ion or electron currents that flow to it in response to different voltages. The current vs. voltage trace, called the *I-V characteristic*, can be analyzed to reveal information about n, T_e, V_s (space potential), and even the distribution function $f_e(v)$, *but not the ion temperature*. Since the probe is immersed in a harsh environment, special techniques are used to protect it from the plasma and vice versa, and to ensure that the circuitry gives the correct $I-V$ values. The probe tip is made of a high-temperature material, usually a tungsten rod or wire 0.1–1 mm in diameter. The rod is threaded into a thin ceramic tube, usually alumina, to insulate it from the plasma except for a short length of exposed tip, about 2–10 mm long. These materials can be exposed to low-temperature laboratory plasmas without melting or excessive sputtering. To avoid disturbing the plasma, the ceramic tube should be as thin as possible, preferably < 1 mm in diameter but usually several times this. The probe tip should extend out of the end of the tube without touching it, so that it would not be in electrical contact with any conducting coating that may deposit onto the insulator. The assembly is encased in a vacuum jacket, which could be a stainless steel or glass tube 1/4″ in outside diameter (OD). It is preferable to make the vacuum seal at the outside end of the probe as-

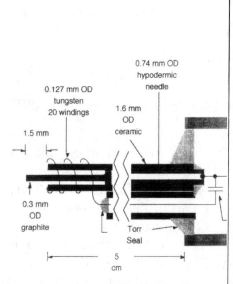

0.127 mm OD
tungsten
20 windings

0.74 mm OD
hypodermic
needle

1.5 mm

1.6 mm
OD
ceramic

0.3 mm
OD
graphite

Torr
Seal

5
cm

Fig. 9. A carbon probe tip assembly with RF compensation circuitry [Sudit and Chen, Plasma Sources Sci. Technol. **4**, 162 (1994)].

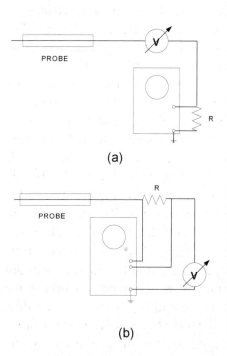

PROBE

R

(a)

R

PROBE

V

(b)

Fig. 10. Two basic configurations for the probe circuit.

sembly rather than at the end immersed in the plasma, which can cause a leak. Only the ceramic part of the housing should be allowed to enter the plasma. Some commercial Langmuir probes use a rather thick metal tube to support the probe tip assembly, and this can modify the plasma characteristics unless the density is very low. In dense plasmas the probe cannot withstand the heat unless the plasma is pulsed or the probe is mechanically moved in and out of the plasma in less than a second. When collecting ion current, the probe can be eroded by sputtering, thus changing its collection area. This can be minimized by using carbon as the tip material. Ordinary pencil lead, 0.3mm in diameter works well and can be supported by a hypodermic needle inside the ceramic shield. One implementation of a probe tip assembly is shown in Fig. 9.

There are two basic ways to apply a voltage V to the probe and measure the current I that it draws from the plasma, and each has its disadvantages. In Fig. 10a, the probe lead, taken through a vacuum feedthru, is connected to a battery or a variable voltage source (*bias supply*) and then to a termination resistor R to ground. To measure the probe current, the voltage across R is recorded or displayed on an oscilloscope. This arrangement has the advantage that the measuring resistor is grounded and therefore not subject to spurious pickup. Since the resistor is usually 10-1000Ω, typically 50Ω, this is not a serious problem anyway. The disadvantage is that the bias supply is floating. If this is a small battery, it cannot easily be varied. If it is a large electronic supply, the capacitance to ground will be so large that ac signals will be short-circuited to ground, and the probe cannot be expected to have good frequency response. The bias supply can also act as an antenna to pick up rf noise. To avoid this, one can ground the bias supply and put the measuring resistor on the hot side, as shown in Fig. 10b. This is usually done if the bias supply generates a sweep voltage. However, the voltage across R now has to be measured with a differential amplifier or some other floating device; or, it can be optoelectronically transmitted to a grounded circuit. The probe voltage V_p should be measured on the ground side of R so as not to load the probe with another stray capacitance.

To measure plasma potential with a Langmuir probe, one can terminate the probe in a high impedance, such as the 1 MΩ input resistance of the oscilloscope. This is called a *floating probe*. A lower R, like 100K, can be used to suppress pickup. The minimum value of

R has to be high enough that the *IR* drop through it does change the measured voltage. A rough rule of thumb is that $I_{sat}R$ should be much greater than T_{eV}, or $R \gg T_{eV}/I_{sat}$, where I_{sat} is the ion saturation current defined below. The voltage measured is not the plasma potential but the floating potential, also defined below. The large value of *R* means that good frequency response is difficult to achieve because of the *RC* time constant of stray capacitances. One can improve the frequency response with *capacitance neutralization* techniques, but even then it is hard to make a floating probe respond to RF frequencies.

2. The electron characteristic

A Langmuir *I-V* trace is usually displayed upside down, so that electron current into the probe is in the +*y* direction. The curve, resembling that in Fig. 11a, has five distinct parts. The point at which the curve crosses the *V* axis is called the *floating potential* V_f. To the left of this the probe draws ion current, and the curve soon flattens out to a more or less constant value called the *ion saturation current* I_{sat}. To the right of V_f, electron current is drawn, and the *I-V* curve goes into an *exponential part*, or *transition region*, as the Coulomb barrier is lowered to allow slower electrons in the Maxwellian distribution to penetrate it. At the space potential V_s, the curve takes a sharp turn, called the *knee*, and saturates at the *electron saturation current* I_{es}. Actual *I – V* curves in RF or magnetized plasmas usually have an indistinct knee, as shown in Fig. 11b.

The exponential part of the *I – V* curve, when plotted semi-logarithmically vs. the probe voltage V_p, should be a straight line if the electrons are Maxwellian:

$$I_e = I_{es} \exp[e(V_p - V_s)/KT_e)], \tag{7}$$

where, from Eq. (A4-2),

$$I_{es} = eAn_e\bar{v}/4 = en_e A \left(\frac{KT_e}{2\pi m}\right)^{1/2}, \tag{8}$$

A being the exposed area of the probe tip. Eq. (7) shows that the slope of the (ln *I*)–V_p curve is exactly $1/T_{eV}$ and is a good measure of the electron temperature. As long as the electrons are Maxwellian and are repelled by the probe, the EEDF at a potential *V* < 0 is proportional to

$$f(v) \propto e^{-(\frac{1}{2}mv^2 + eV)/KT_e} = e^{-e|V|/KT_e}e^{-(mv^2/2KT_e)}. \tag{9}$$

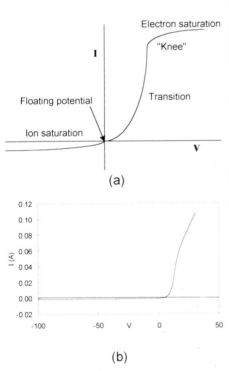

(a)

(b)

Fig. 11. (a) An idealized *I – V* characteristic showing its various parts; (b) a real *I – V* curve from an ICP.

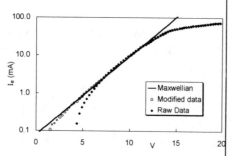

Fig. 12. A semilog plot of electron current from an *I − V* curve in an ICP

We see that *f*(v) is still Maxwellian at the same T_e; only the density is decreased by $\exp(-e|V|/KT_e)$. Thus, the slope of the semilog curve is independent of probe area or shape and independent of collisions, since these merely preserve the Maxwellian distribution. However, before I_e can be obtained from *I*, one has to subtract the ion current I_i. This can be done approximately by drawing a straight line through I_{sat} and extrapolating it to the electron region.

One can estimate the ion contribution more accurately by using one of the theories of ion collection discussed below, but refinements to this small correction are usually not necessary, and they affect only the high-energy tail of the electron distribution. One easy iteration is to change the magnitude of the I_{sat} correction until the ln *I* plot is linear over as large a voltage range as possible. Fig. 12 shows a measured electron characteristic and a straight-line fit to it. The ion current was calculated from a theoretical fit to I_{sat} and added back to *I* to get I_e. The uncorrected points are also shown; they have a smaller region of linearity.

3. Electron saturation

Since I_{sat} is $\approx c_s$ because of the Bohm sheath criterion, I_{es}, given by Eq. (8), should be $\approx (M/m)^{1/2}$ times as large as I_{sat}. In low-pressure, unmagnetized discharges, this is indeed true, and the knee of the curve is sharp and is a good measure of V_s. For very high positive voltages, I_{es} increases as the sheath expands, the shape of the curve depending on the shape of the probe tip. However, effects such as collisions and magnetic fields will lower the magnitude of I_{es} and round off the knee so that V_s is hard to determine. In particular, magnetic fields strong enough to make the electron Larmor radius smaller than the probe radius will limit I_{es} to only 10-20 times I_{sat} because the probe depletes the field lines that it intercepts, and further electrons can be collected only if they diffuse across the B-field. The knee, now indistinct, indicates a space potential, but only that in the depleted tube of field lines, not V_s in the main plasma. In this case, the *I − V* curve is exponential only over a range of a few KT_e above the floating potential and therefore samples only the electrons in the tail of the Maxwellian. One might think that measurement of I_{es} would give information on the electron density, but this is possible only at low densities and pressures, where the mean free path is very long. Otherwise, the current collected by the probe is so large that it drains the plasma and changes its equilibrium properties. It is better to measure n by collecting ions, which would give the same information, since plasmas are quasineutral. More importantly, one should avoid collecting saturation electron current for more than a few milliseconds at a time, because the probe can be damaged.

4. Space potential

The time-honored way to obtain the space potential (or plasma potential) is to draw straight lines through the $I - V$ curve in the transition and electron saturation regions and call the crossing point V_s, I_{es}. This does not work well if I_{es} region is curved. As seen in Fig. 11b, a good knee is not always obtained even in an ICP with $\mathbf{B}_0 = 0$. In that case, there are two methods one can use. The first is to measure V_f and calculate it from Eq. (A4-4), regarding the probe as a wall. The second is to take the point where I_e starts to deviate from exponential growth; that is, where $I'_e(V)$ is maximum or $I''_e(V)$ is zero. If $I'_e(V)$ has a distinct maximum, a reasonable value for V_s is obtained, but it would be dangerous to equate the current there to I_{es}. That is because, according to Eq. (7), I_{es} depends exponentially on the assumed value of V_s.

5. Ion saturation current[1]

a) Plane probes. The measurement of I_{sat} is the simplest and best way to determine n. At densities above about 10^{11} cm^{-3}, the sheath around a negatively biased probe is so thin that the area of the sheath edge is essentially the same as the area of the probe tip itself. The ion current is then just that necessary to satisfy the Bohm sheath criterion:

$$I_{sat} = 0.5eAn(KT_e / M)^{1/2} , \qquad (10)$$

where the factor 0.5 represents n_s/n. This value is only approximate; when probes are calibrated against other diagnostics, such as microwave interferometry, a factor of 0.6-0.7 has been found to be more accurate. Note that Eq. (10) predicts a constant I_{sat}, which can happen only for flat probes in which the sheath area cannot expand as the probe is made more and more negative. In practice, I_{sat} usually has a slope to it. This is because the ion current has to come from a disturbed volume of plasma (the presheath) where the ion distribution changes from iso-

[1] For detailed references, see F.F. Chen, *Electric Probes*, in "Plasma Diagnostic Techniques", ed. by R.H. Huddlestone and S.L. Leonard (Academic Press, New York, 1965), Chap. 4, pp. 113-200.

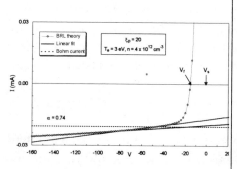

Fig. 13. Illustrating the extrapolation of I_i back to the floating potential to get I_{sat}. In this case, the Bohm coefficient 0.5 in Eq. (10) has to be replaced by 0.74 to get the right density.

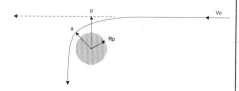

Fig. 14. Definition of impact parameter p.

tropic to unidirectional. If the probe is a disk of radius R, say, the disturbed volume may have a size comparable to R, and would increase as the $|V_p|$ increases. In that case, one can extrapolate I_i back to V_f to get a better measure of I_{sat} before the expansion of the presheath. This is illustrated in Fig. 13. Better saturation with a plane probe can be obtained by using a *guard ring*, a flat washer-shaped disk surrounding the probe but not touching it. It is biased at the same potential as the probe to keep the fields planar as V_p is varied. The current to the guard ring is disregarded. A section of the chamber wall can be isolated to be used as a plane probe with a large guard ring.

b) Cylindrical probes

i) OML theory. As the negative bias on a probe is increased to draw I_i, the sheath on cylindrical and spherical probes expands, and I_i does not saturate. Fortunately, the sheath fields fall off rapidly away from the probe so that exact solutions for $I_i(V_p)$ can be found. We consider cylindrical probes here because spherical ones are impractical to make, though the theory for them converges better. The simplest theory is the orbital-motion-limited (OML) theory of Langmuir.

Consider ions coming to the attracting probe from infinity in one direction with velocity v_0 and various impact parameters p. The plasma potential V is 0 at ∞ and is negative everywhere, varying gently toward the negative probe potential V_p. Conservation of energy and angular momentum give

$$\tfrac{1}{2}\,mv_0^2 = \tfrac{1}{2}\,mv_a^2 + eV_a \equiv -eV_0$$
$$pv_0 = av_a \tag{11}$$

where $eV < 0$ and a is the distance of closest approach to the probe of radius R_p. Solving, we obtain

$$\tfrac{1}{2}mv_a^2 = \tfrac{1}{2}mv_0^2\left(1+\frac{V_a}{V_0}\right), \qquad p = a\frac{v_a}{v_0} = a\left(1+\frac{V_a}{V_0}\right)^{1/2}. \tag{12}$$

If $a \leq R_p$, the ion is collected; thus, the *effective* probe radius is $p(R_p)$. For monoenergetic particles, the flux to a

[2] F.F. Chen, Phys. Plasmas **8**, 3029 (2001).
[3] F.F. Chen, J.D. Evans, and D. Arnush, Phys. Plasmas **9**, 1449 (2002)
[4] I.D. Sudit and F.F. Chen, *RF compensated probes for high-density discharges*, Plasma Sources Sci. Technol. **3**, 162 (1994).
[5] N. Hershkowitz, *How Langmuir Probes Work*, in Plasma Diagnostics, Vol. 1, Ed. by O. Auciello and D.L. Flamm (Acad. Press, N.Y., 1994), Chap. 3, p. 113.

probe of length L is therefore

$$\Gamma = 2\pi R_p L (1 + V_a / V_0)^{1/2} \Gamma_r , \qquad (13)$$

where Γ_r is the random flux of ions of that energy. Langmuir then extended this result to energy distributions which were Maxwellian at some large distance $r = s$ from the probe, where s is the "sheath edge". The random flux Γ_r is then given by the usual formula

$$\Gamma_r = n \left(\frac{KT_i}{2\pi M} \right)^{1/2} . \qquad (14)$$

With A_p defined as the probe area, integrating over all velocities yields the cumbersome expression

$$\Gamma = A_p \Gamma_r \left\{ \frac{s}{a} \mathrm{erf}(\Phi^{1/2}) + e^\chi [1 - \mathrm{erf}(\chi + \Phi)^{1/2}] \right\} , \qquad (15)$$

where $\quad \chi \equiv -eV_p / KT_i, \quad \Phi \equiv \left(\dfrac{a^2}{s^2 - a^2} \right)\chi, \quad a = R_p .$

Fortunately, there are small factors. In the limit $s \gg a$, when OML theory applies, if at all, we have $\Phi \ll \chi$, and for $T_i \to 0$, $1/\chi \ll 1$. Expanding in Taylor series, we find that the T_i dependences of χ and Γ_r cancel, and a finite limiting value of the OML current exists, independently of the value of T_i.

$$I \xrightarrow[T_i \to 0]{} A_p n e \frac{\sqrt{2}}{\pi} \left(\frac{|eV_p|}{M} \right)^{1/2} . \qquad (16)$$

Thus, the OML current is proportional to $|V_p|^{1/2}$, and the $I - V$ curve is a parabola, while the $I^2 - V$ curve is a straight line. This scaling is the result of conservation of energy and angular momentum. Because ions have large angular momentum at large distances, though they have small velocities, they tend to orbit the probe and miss it. The probe voltage draws them in. The value of T_i cancels out mathematically, but T_i has to be finite for this physical mechanism to work.

The OML result, though simple, is very restricted in applicability. Since the sheath radius s was taken to be infinite, the density has to be so low that the sheath is much larger than the probe. The potential variation $V(r)$ has to be gentle enough that there does not exist an "absorption radius" inside of which the E-field is so strong that no ions can escape being collected. Except in very tenuous plasmas, a well developed sheath and an absorption radius exist, and OML theory is inapplicable. Nonetheless, the $I^2 - V$ dependence of I_{sat} is often observed and is mistakenly taken as evidence of orbital motion.

ii) ABR theory. To do a proper sheath theory, one has to solve Poisson's equation for the potential $V(r)$ everywhere from the probe surface to $r = \infty$. Allen, Boyd, and Reynolds (ABR) simplified the problem by assuming *ab initio* that $T_i = 0$, so that there are no orbital motions at all: the ions are all drawn radially into the probe. Originally, the ABR theory was only for spherical probes, but it was later extended to cylindrical probes by Chen[1], as follows. Assume that the probe is centered at $r = 0$ and that the ions start at rest from $r = \infty$, where $V = 0$. Poisson's equation in cylindrical coordinates is

$$\frac{1}{r}\frac{\partial}{\partial r}\left(r\frac{\partial V}{\partial r}\right) = \frac{e}{\varepsilon_0}(n_e - n_i), \qquad n_e = n_0 e^{eV/KT_e}. \quad (17)$$

To electrons are assumed to be Maxwellian. To find n_i, let I be the total ion flux per unit length collected by the probe. By current continuity, the flux per unit length at any radius r is

$$\Gamma = n_i v_i = I/2\pi r, \quad \text{where} \quad v_i = \left(-2eV/M\right)^{\frac{1}{2}}. \quad (18)$$

Thus,
$$n_i = \frac{\Gamma}{v_i} = \frac{I}{2\pi r}\left(\frac{-2eV}{M}\right)^{-1/2}. \quad (19)$$

Poisson's equation can then be written

$$\frac{1}{r}\frac{\partial}{\partial r}\left(r\frac{\partial V}{\partial r}\right) = -\frac{e}{\varepsilon_0}\left[\frac{I}{2\pi r}\left(\frac{-2eV}{M}\right)^{-1/2} - n_0 e^{eV/KT_e}\right] \quad (20)$$

Defining
$$\eta \equiv -\frac{eV}{KT_e}, \qquad c_s \equiv \left(\frac{KT_e}{M}\right)^{1/2}, \quad (21)$$

we can write this as

$$\frac{KT_e}{e}\frac{1}{r}\frac{\partial}{\partial r}\left(r\frac{\partial \eta}{\partial r}\right) = -\frac{e}{\varepsilon_0}\left[\frac{I}{2\pi r}\frac{(2\eta)^{-1/2}}{c_s} - n_0 e^{-\eta}\right] \quad (22)$$

or

$$-\frac{\varepsilon_0 KT_e}{n_0 e^2}\frac{1}{r}\frac{\partial}{\partial r}\left(r\frac{\partial \eta}{\partial r}\right) = \frac{I}{2\pi r}\frac{(2\eta)^{-1/2}}{n_0 c_s} - e^{-\eta}. \quad (23)$$

The Debye length appears on the left-hand side as the natural length for this equation. We therefore normalize r to λ_D by defining a new variable ξ:

$$\xi \equiv \frac{r}{\lambda_D}, \qquad \lambda_D \equiv \left(\frac{\varepsilon_0 K T_e}{n_0 e^2}\right)^{1/2}. \qquad (24)$$

Eq. (23) now becomes

$$\frac{\partial}{\partial \xi}\left(\xi \frac{\partial \eta}{\partial \xi}\right) = \frac{I\xi}{2\pi r}\frac{1}{n_0 c_s}(2\eta)^{-1/2} - \xi e^{-\eta}$$

$$= \frac{I}{2\pi n_0}\frac{1}{\lambda_D c_s}(2\eta)^{-1/2} - \xi e^{-\eta}$$

$$= \frac{I}{2\pi n_0}\left(\frac{n_0 e^2}{\varepsilon_0 K T_e}\frac{M}{K T_e}\right)^{1/2}(2\eta)^{-1/2} - \xi e^{-\eta}$$

$$= \frac{eI}{2\pi K T_e}\left(\frac{M}{2\varepsilon_0 n_0}\right)^{1/2}\eta^{-1/2} - \xi e^{-\eta}$$

$$(25)$$

Defining

$$J \equiv \frac{eI}{2\pi K T_e}\left(\frac{M}{2\varepsilon_0 n_0}\right)^{1/2}, \qquad (26)$$

we arrive at the ABR equation for cylindrical probes:

$$\frac{\partial}{\partial \xi}\left(\xi \frac{\partial \eta}{\partial \xi}\right) = J\eta^{-1/2} - \xi e^{-\eta}. \qquad (27)$$

For each assumed value of J (normalized probe current), this equation can be integrated from $\xi = \infty$ to any arbitrarily small ξ. The point on the curve where $\xi = \xi_p$ (the probe radius) gives the probe potential η_p for that value of J. By computing a family of curves for different J (Fig. 15), one can obtain a $J - \eta_p$ curve for a probe of radius ξ_p by cross-plotting (Fig 16). Of course, both J and ξ_p depend on the unknown density n_0, which one is trying to determine from the measured current I_i. ($K T_e$ is supposed to be known from the electron characteristic.) The extraction of n_0 from these universal curves is a trivial matter for a computer. In the graphs the quantity $J\xi_p$ is plotted, since that is independent of n_0. Note that for small values of ξ_p, I^2 varies linearly with V_p, as in OML theory, but for entirely different reasons, since there is no orbiting here.

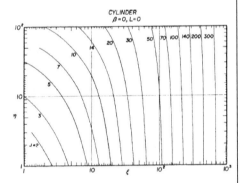

Fig. 15. ABR curves for $\eta(\xi)$.

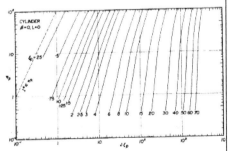

Fig. 16. *V–I* curves derived from $\eta(\xi)$.

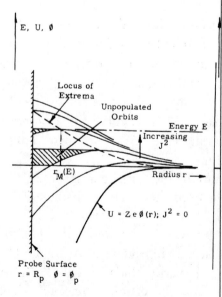

Fig. 17. Definition of absorption radius.

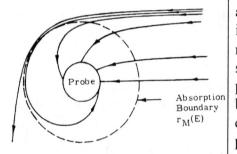

Fig. 18. Effective potential seen by ions with angular momentum J.

iii) BRL theory. The first probe theory which accounted for both sheath formation and orbital motions was published by Bernstein and Rabinowitz (BR), who assumed an isotropic distribution of ions of a single energy E_i. This was further refined by Laframboise (L), who extended the calculations to a Maxwellian ion distribution at temperature T_i. The BRL treatment is considerably more complicated than the ABR theory. In ABR, all ions strike the probe, so the flux at any radius depends on the conditions at infinity, regardless of the probe radius. That is why there is a set of universal curves. In BRL theory, however, the probe radius must be specified beforehand, since those ions that orbit the probe will contribute twice to the ion density at any given radius r, while those that are collected contribute only once. The ion density must be known before Poisson's equation can be solved, and clearly this depends on the presence of the probe. There is an "absorption radius" (Fig. 17), depending on J, inside of which all ions are collected. Bernstein solved the problem by expressing the ion distribution in terms of energy E and angular momentum J instead of v_r and $v_\perp$. Ions with a given J see an effective potential barrier between them and the probe. They must have enough energy to surmount this barrier before they can be collected. In Fig. 18, the lowest curve is for ions with $J = 0$; these simply fall into the probe. Ions with finite J see a potential hill. With sufficient energy, they can climb the hill and fall to the probe on the other side. The dashed line through the maxima shows the absorption radius for various values of J.

The computation tricky and tedious. It turns out that KT_i makes little difference if $T_i/T_e < 0.1$ or so, as it usually is. Laframboise's extension to a Maxwellian ion distribution is not normally necessary; nonetheless, Laframboise gives the most complete results. Fig. 19 shows an example of ion saturation curves from the BRL theory. One sees that for large probes ($R_p/\lambda_D \gg 1$) the ion current saturates well, since the sheath is thin. For small R_p/λ_D, I_i grows with increasing V_p as the sheath radius increases.

One might think that the ABR result would be recovered if takes $T_i = 0$ or $E_i = 0$ in the BRL computation. However, this happens only for spherical probes. For cylindrical probes, there is a problem of nonuniform convergence. Since the angular momentum is Mvr, for $r \to \infty$ ions with zero thermal velocity have $J = (M)(0)(\infty)$, an indeterminate form. The correct treatment is to calculate the probe current for $T_i > 0$ and *then* take the limit T_i

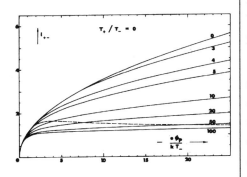

Fig. 19. Laframboise curves for $I_i - V$ characteristics in dimensionless units, in the limit of cold ions. Each curve is for a different ratio R_p/λ_D.

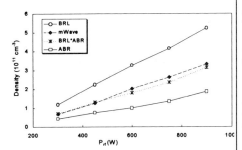

Fig. 20. Comparison of n measured with microwaves with probes using two different probe theories.

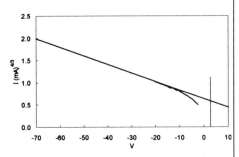

Fig. 21. Extrapolation to get I_i at V_f.

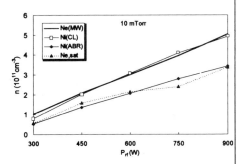

Fig. 22. Comparison of microwave and probe densities using the floating potential method (CL), ABR theory, and $I_{e,sat}$.

→ 0, as BRL have done. The BRL predictions have been borne out in experiments in fully ionized plasmas, but not in partially ionized ones.

iv) Comparison among theories. It is not reasonable to reproduce the ABR or BRL computations each time one makes a probe measurement. Chen[2] has solved this problem by parametrizing the ABR and BRL curves so that the $I_i - V$ curve can be easily be created for any value of R_p/λ_D. One can then compute the plasma density from the probe data using the ABR and BRL theories and compare with the density measured with microwave interferometry. Such a comparison is shown in Fig. 20. One sees that the ABR theory predicts too low a density because orbiting is neglected, and therefore the predicted current is too high and the measured current is identified with a lower density. Conversely, BRL theory predicts too high a density because it assumes more orbiting than actually occurs, so that the measured current is identified with a high density. This effect occurs in partially ionized plasmas because the ions suffer charge-exchange collisions far from the probe, outside the sheath, thus losing their angular momentum. The BRL theory assumes that the ions retain their angular momentum all the way in from infinity. One might expect the real density to lie in between, and indeed, it agrees quite well with the geometric mean of the BRL and ABR densities.

Treating the charge-exchange collisions rigorously in the presheath would be an immense problem, but recently Chen et al.[3] have found an even easier, fortuitous, way to estimate the plasma density in ICPs and other processing discharges. The method relies on finding the ion current at floating potential V_f by extrapolating on a graph of $I_i^{4/3}$ vs. V_p, as shown in Fig. 21. The power 4/3 is chosen because it usually leads to a straight line graph. At $V_p = V_f$, let the sheath thickness d be given by the Child-Langmuir formula of Eq. (A4-7) with $V_0 = V_f$. The sheath area is then $A = 2\pi(R_p+d)L$. If the ions enter the sheath at velocity c_s with density $n_s = \frac{1}{2}n_0c_s$, the ion current is $I_i = \frac{1}{2}n_0eAc_s$, and n can easily be calculated from the extrapolated value of $I_i(V_f)$. Note that if $R_p \ll d$, Eq. (A4-7) predicts the $I_i^{4/3} - V_p$ dependence (but this is accidental). (Since $d \propto \lambda_D \propto n^{-1/2}$, n has to be found by iteration or by solving a quadratic equation.) The density in a 10 mTorr ICP discharge is shown in Fig. 22, compared to n measured by microwaves, by probes using the ABR theory, or by probes using the saturation electron current. The V_f(CL) method fits best, though the fit is not always this good. The OML theory (not shown) also fits poorly. Though this is a fast and easy method to interpret I_{sat} curves, it is hard to justify because the CL formula of Eq. (A4-7) applies to planes, not cylinders, and the Debye sheath thickness has been neglected, as well as orbiting and collisions. This simple-minded approach apparently works because the neglected effects cancel one another. From the preceding discussion, it is clear that the rigorous theories, ABR and BRL, can err by a factor of 2 or more in the value of *n* in partially ionized plasmas. There are heuristic methods, but these may not work in all conditions. It is difficult for Langmuir probes to give a value of *n* accurate to better than 10–20%; fortunately, such accuracy is not often required.

6. Distribution functions

Since the ion current is insensitive to T_i, Langmuir probes cannot measure ion temperature, and certainly not the ion velocity distribution. However, careful measurement of the transition region of the $I - V$ characteristic can reveal the electron distribution if it is isotropic. If the probe surface is a plane perpendicular to *x*, the electron flux entering the sheath depends only on the *x* component of velocity, v_x. For instance, the Maxwell distribution for v_x is

$$f_M(v_x) = \left(\frac{1}{v_{th}\sqrt{\pi}}\right)\exp(-v_x^2/v_{th}^2), \quad v_{th}^2 \equiv 2KT_e/m.$$

(28)

The coefficient normalizes $f(v)$ so that its integral over all v_x's is unity. If $f(v)$ is not Maxwellian, it will have another form and another coefficient in front. The electron current that can get over the Coulomb barrier and be collected by the probe will therefore be

$$I_e = eAn\int_{v_{min}}^{\infty} v_x f(v_x)dv_x, \quad \tfrac{1}{2}mv_{min}^2 = e(V_s - V_p) = -eV_p$$

(29)

where v_{min} is the minimum energy of an electron that can reach the probe, and $V_s = 0$ by definition. Taking the derivative and simplifying, we find

$$\frac{dI_e}{dV_p} = eAn\frac{d}{dV_p}\int_{v_{min}}^{\infty} v_x f(v_x)\frac{dv_x}{dV_p}dV_p$$

$$= -eAn\, v_x f(v_x)\frac{dv_x}{dV_p}\Bigg|_{v_x = v_{min}}$$

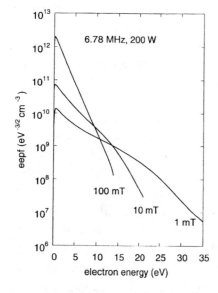

Fig. 23. EEDF curves obtained with a Langmuir probe in a TCP discharge [Godyak et al. J. Appl. Phys. **85**, 3081, (1999)].

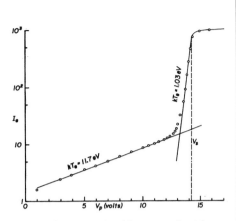

Fig. 24. An *I – V* curve of a bi-Maxwellian EEDF.

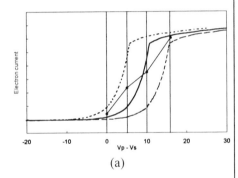

(a)

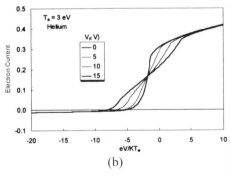

(b)

Fig. 25 . (a) The center curve is the correct *I – V* curve. The dashed ones are displaced by ±5V, representing changes in V_s. At the vertical lines, the average I_e between the displaced curves is shown by the dot. The line through the dots is the time-averaged *I – V* curve that would be observed, differing greatly from the correct curve. (b) Computed *I – V* curves for sinusoidal V_s oscillations of various amplitudes.

$$\frac{dI_e}{dV_p} = -eAn v_x f(v_x) \frac{-e}{m v_x}\bigg|_{v_x = v_{min}} = A\frac{ne^2}{m} f(v_{min}),(30)$$

so that $f(v_x)$ can be found from the first derivative of the *I – V* curve. If the probe is not flat, however, one has to take the three-dimensional distribution $g(v) = 4\pi v^2 f(v)$, where v is the absolute value |v| of the velocity, and take into account the various angles if incidence. Without going into the details, we then find, surprisingly, that $f(v)$ is proportional to the *second* derivative of the *I – V* curve:

$$\frac{d^2 I_e}{dV_p^2} \propto f(v) \qquad (31)$$

This result is valid for any convex probe shape as long as the distribution is isotropic, and for any anisotropic distribution if the probe is spherical. To differentiate *I – V* data twice will yield noisy results unless a good deal of smoothing is employed. Alternatively, one can *dither* the probe voltage by modulating it at a low frequency, and the signal at the dither frequency will be proportional to the first derivative. In that case, only one further derivative has to be taken to get $f(v)$. Figure 23 is an example of non-Maxwellian $f(v)$'s obtained by double differentiation with digital filtering.

   In special cases where the EEDF consists of two Maxwellians with well separated temperatures, the two KT_e's can be obtained by straight-line fits on the semilog *I – V* curve without complicated analysis. An example of this is shown in Fig. 24.

7. RF compensation

   Langmuir probes used in RF plasma sources are subject to RF pickup which can greatly distort the *I – V* characteristic and give erroneous results. ECR sources which operate in the microwave regime do not have this trouble because the frequency is so high that it is completely decoupled from the circuitry, and the measured currents are the same as in a DC discharge. However, in RF plasmas, the space potential can fluctuate is such a way that the circuitry responds incorrectly. The problem is that the *I – V* characteristic is nonlinear. The "V" is actually the potential difference $V_p - V_s$, where V_p is a DC potential applied to the probe, and V_s is a potential that can fluctuate at the RF frequency and its harmonics. If one displaces the *I – V* curve horizontally back and

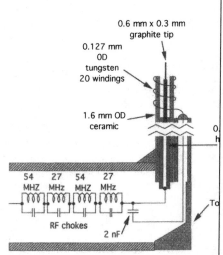

Fig. 26. Design of a dogleg probe.

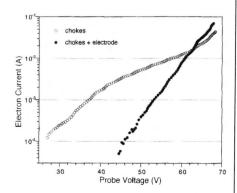

Fig. 27. *I – V* curves taken with and without an auxiliary electrode.

forth around a center value V_0, the average current I measured will not be $I(V_0)$, since I varies exponentially in the transition region and also changes slope rapidly as it enters the ion and electron saturation regions. The effect of this is to make the $I - V$ curve wider, leading to a falsely high value of T_e and shifting the floating potential V_f to a more negative value. This is illustrated in Fig. 25.

Several methods are available to correct for this. One is to tap off a sinusoidal RF signal from the power supply and mix this with the probe signal with variable phase and amplitude. When the resultant $I - V$ curve gives the lowest value of T_e, one has probably simulated the V_s oscillations. This method has the disadvantage that the V_s oscillations can contain more than one harmonic. A second method is to measure the V_s oscillations with another probe or section of the wall which is floating, and add that signal to the probe current signal with variable phase and amplitude. The problem with this method is that the V_s fluctuations are generally not the same everywhere. A third method is to isolate the probe tip from the rest of the circuit with an RF choke (inductor), so that the probe tip is floating at RF frequencies but is fixed at the DC probe bias at low frequencies. The problem is that the probe tip does not draw enough current to fill the stray capacitances that connect it to ground at RF frequencies. One way is to place a large slug of metal inside the insulator between the probe tip and the chokes. This metal slug has a large area and therefore picks up enough charge from the V_s oscillations to drive the probe tip to follow them. However, we have found[4] that the best way is to use an external floating electrode, which could be a few turns of wire around the probe insulator, and connect it through a capacitor to a point between the probe tip and the chokes. The charge collected by this comparatively large "probe" is then sufficient to drive the probe tip so that V_p - V_s remains constant. Note that this auxiliary electrode supplies only the RF voltage; the dc part is still supplied by the external power supply. The design of the chokes is also critical: they must have high enough Q to present a resonantly high impedance at both the fundamental and the second harmonic of the RF frequency. This is the reason there are two pairs of chokes in Fig. 9. One pair is resonant at ω, and the other at 2ω. Two chokes are used in series to increase the Q. A compromise has to be made between high Q and small physical size of the chokes. Figure 26 shows a "dogleg" design, which permits scans in two directions. Figure 27 shows an $I - V$ curve taken with and without the auxiliary electrode, showing that the chokes

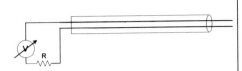

Fig. 28. A double probe.

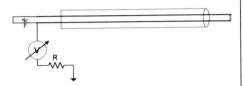

Fig. 29. A hot probe.

themselves are usually insufficient.

Without proper RF compensation, Langmuir probe data in RF discharges can give spurious data on T_e, V_f, and $f(v)$. However, if one needs to find only the plasma density, the probe can be biased so that V never leaves the ion saturation region, which is linear enough that the average I_{sat} will be the correct value.

8. Double probes and hot probes[5]

When V_s fluctuates slowly, one can use the method of double probes, in which two identical probes are inserted into the plasma in close proximity, and the current from one to the other is measured as a function of the voltage difference between them. The $I - V$ characteristic is then symmetrical and limited to the region between the I_{sat}'s on each probe. If the probe array floats up and down with the RF oscillations, the $I - V$ curve should not be distorted. This method does not work well in RF plasmas because it is almost impossible to make the whole two-probe system float at RF frequencies because of the large stray capacitance to ground. Even if both tips are RF compensated, the RF impedances must be identical.

Hot probes are small filaments that can be heated to emit electrons. These electrons, which have very low energies corresponding to the KT of the filament, cannot leave the probe as long as $V_p - V_s$ is positive. As soon as $V_p - V_s$ goes negative, however, the thermionic current leaves the probe, and the probe current is dominated by this rather than by the ion current. Where the $I - V$ curve crosses the x axis, therefore, is a good measure of V_s. The voltage applied to the filament to heat it can be eliminated by turning it off and taking the probe data before the filament cools. One can also heat the probe by bombarding it with ions at a very large negative V_p, and then switching this voltage off before the measurement. In general it is tricky to make hot probes small enough. For further information on these techniques and on behavior of Langmuir probes in RF plasmas, the reader is referred to the chapter by Hershkowitz[5].

XIII. OTHER LOCAL DIAGNOSTICS

1. Magnetic probes

a) Principle of operation. Fluctuating RF magnetic fields inside the plasma can be measured with a magnetic probe, which is a small coil of wire, perhaps 2 mm in diameter, covered with glass or ceramic so as to protect it from direct exposure to charged particles.

When the coil is placed in a time-varying magnetic field **B**, an electric field is induced along the wire according to Faraday's Law:

$$\nabla \times \mathbf{E} = -d\mathbf{B}/dt . \tag{32}$$

Integrating this over the surface enclosed by the coil with the help of Stokes' theorem to convert the surface integral to a line integral, we obtain

$$\int \dot{\mathbf{B}} \cdot dS \equiv \dot{\Phi} = -\int (\nabla \times \mathbf{E}) \cdot dS = -\int \mathbf{E} \cdot d\ell \equiv -V_{ind} . \tag{33}$$

Here the line integral is along the wire in the coil and Φ is the magnetic flux through the coil, which is $\approx BA$, where A is the area of the coil. The induced voltage V_{ind} is measured by a high-impedance device like an oscilloscope. If there are N turns in the coil, the voltage will be N times higher; hence,

$$V_{ind} = -NA\dot{B} \tag{34}$$

The dot indicates the time-derivative and is the origin of the name "B-dot probe." The minus sign indicates that the induced electric field is in the opposite direction to that obtained when the right-hand rule is applied to **B**. For a sinusoidal signal, B-dot is proportional to ωB, so that the probe is more sensitive to higher frequencies. To obtain B from the measured V_{ind}, one can use a simple integrator consisting of a resistor and a capacitor to ground to obtain

$$B = -\frac{1}{NA} \int V_{ind} dt . \tag{35}$$

One only has to be sure that the RC time constant of the integrator is much longer than the period of the signal.

b) Construction. The probe itself can be as simple as ten turns of thin wire wound on a core machined out of boron nitride. The coil can be placed inside a ceramic tube or a closed glass tube. Such a tube is necessarily larger than a Langmuir probe shaft and may disturb the plasma downstream from the source. If the axis of the coil is parallel to the tube, the component of **B** parallel to the shaft will be measured. If the coil axis is perpendicular to the shaft, one can change from B_r to B_θ measurement by rotating the shaft by 90°. Sometimes three coils are mounted in the same shaft to measure all three B components at the same time.

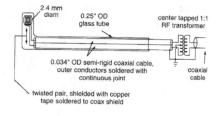

Fig. 30. A magnetic probe with a balun transformer.

The difficult part is to take the signal out through the probe shaft without engendering too much RF pickup. One way is to use a very thin rigid coax, which is then connected to the scope with a 50-Ω cable. The coil in this case can be a single turn formed from the center conductor looped around and soldered to the conducting shield. If the shaft has to traverse a long path through the plasma, a better way is to use a multi-turn coil to increase the signal voltage, and then bring the two ends of the coils through the shaft with a twisted pair of wires. Outside the plasma, the wires are connected to a balun (balanced-to-unbalanced) 1-to-1 transformer so that the signal can be carried to the scope with an unbalanced line. Such a probe is shown in Fig. 30. The transformer can also have a turns ratio that amplifies the signal voltage. With magnetic probes there is always the danger of capacitive pickup through the insulators. One can check this by rotating the probe 180°. The magnetic signal should be the same in magnitude but shifted 180° in phase, while the capacitive signal would be the same in both orientations. Whether or not the probe and leads should be shielded with slotted conductors is a matter of experimentation; the shield can help or actually make the pickup worse.

2. Energy analyzers

Gridded energy analyzers are used to obtain better data for ion and electron energy distributions than can be obtained with Langmuir probes. However, these instruments are necessarily large—at least 1 cubic centimeter in volume—and will disturb the plasma downstream from them. A standard gridded analyzer has four grids: 1) a grounded or floating outer grid to isolate the analyzer from the plasma, 2) a grid with positive or negative potential to repel the unwanted species, 3) a solid collector with variable potential connected to the current measuring device, and 4) a suppressor grid in front of the collector to repel secondary electrons. In Fig. 31, *s* is the sheath edge. Grid G1, whether floating or grounded, will be negative with respect to the plasma, and therefore will repel all electrons except the most energetic ones. One cannot bias this grid positively, since it will then draw so much electron current that the plasma will be disturbed. It is sometime omitted in order to allow slower electrons to enter the analyzer. Grid G1 also serves to attenuate the flux of plasma into the analyzer so that the Debye length is not so short there that subsequent grid wires will be shielded out. In the space be-

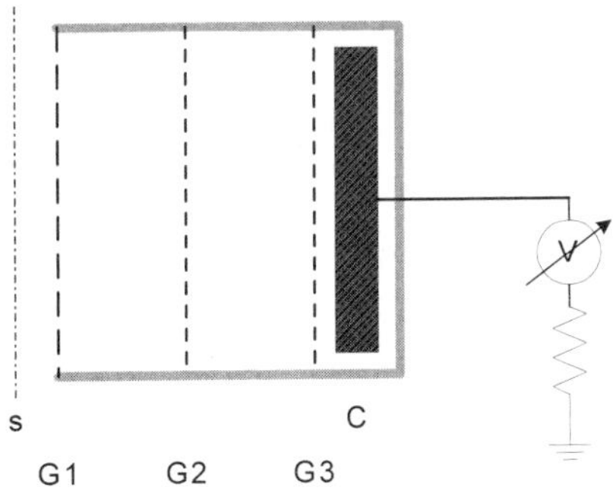

Fig. 31. A gridded energy analyzer.

hind Grid G1, there will be a distribution of ions which have been accelerated by the sheath but which still has the original relative energy distribution (unless it has been degraded by scattering off the grid wires). These are neutralized by electrons that have also come through G1. These electrons also have the original relative energy distribution, but they all have been decelerated by the sheath. Grid G2 is set positive to repel ions and negative to repel electrons. For example, to obtain $f_i(v)$, we would set G2 sufficiently negative (V_2) to repel all the electrons. The ions will then be further accelerated toward the collector. This collecting plate C, at V_c, would collect all the ion current if it were at the same potential as V_2. By biasing it more and more positive relative to V_2, only the most energetic ions would be collected. The curve of I vs. V_c would then give $f_i(v)$ when it is differentiated. When ions strike the collector, secondary electrons can be emitted, and these will be accelerated away from the collector by the field between C and G2, leading to a false enhancement of the apparent ion current. To prevent this, Grid G3 is fixed at a small negative potential (about $-2V$) relative to C) so that these electrons are turned back. Variations to this standard configuration are also possible.

In a plasma with RF fluctuations, energy analyzers would suffer from nonlinear averaging, just as Langmuir probes do. Because of their large size, and therefore stray capacitance, it would not be practical to drive the grids of an energy analyzer to follow changes in plasma potential at the RF frequency. However, one can design the circuitry to be fast enough to follow the RF and then record the oscillations in collected current as a function of time during each RF cycle. By selecting data from the same RF phase to perform the analysis, one can, in principle, obtain the true energy distribution. This technique cannot be used for Langmuir probes, because the currents there are so small that the required frequency response cannot be obtained. RF-sensitive energy analyzers have been made successfully by at least two groups; one such analyzer is shown in Fig. 32.

3. RF current probe

Current probes, sometimes called Rogowski coils, are coils of wire wound on a toroidal coil form shaped like a Life Saver. Figure 33 shows such a coil. Current passing through the hole induces a magnetic field in the azimuthal direction, and this, in turn, induces a voltage in the turns of wire. The current driven through the wire is

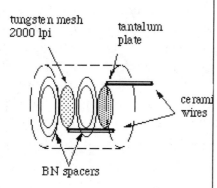

tungsten mesh
2000 lpi

tantalum plate

ceramic wires

BN spacers

Fig. 32. Energy analyzer with only one grid and a collector

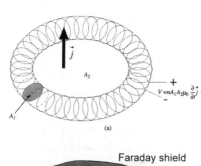

$\vec{j}$

A_2

$V = n A_1 A_2 \mu_0 \frac{\partial \vec{j}}{\partial t} \cdot \hat{n}$

A_1

(a)

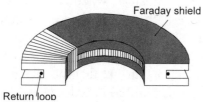

Faraday shield

Return loop

Fig. 33. Construction of an RF current probe.

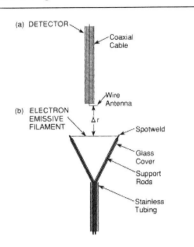

Fig. 34. Schematic of a POP [Shirakawa and Sugai, Jpn. J. Appl. Phys. **32**, 5129 (1993)].

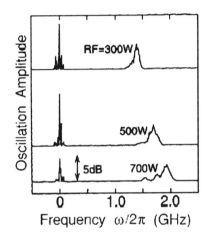

Fig. 35. Peak at ω_p (at the right) moves as P_{rf} is increased [*ibid.*].

then measured in an external circuit. The coil must take a return loop the long way around the torus to cancel the B-dot pickup that is induced by B-fields that thread the hole. Current probes are usually large and can be bought as attachments to an oscilloscope, but these are unsuitable for insertion into a plasma. The probe shown here is not only small (~1 cm diam) but is also made for RF frequencies. It is covered with a Faraday shield to reduce electrostatic pickup, and the windings are carefully calibrated so that the B-dot and E-dot signals are small compared with the J-dot signal. An example of a J-dot measurement was shown in Fig A6-17.

4. Plasma oscillation probe

When used in a plasma processing reactor, Langmuir probes tend to get covered with insulating coatings so that they can no longer properly measure dc current. A plasma oscillation probe avoids this by measuring only ac signals, which can pass capacitively through the coatings. A filament, like a hot probe (Fig. 34), is heated to emission and biased to ~100V negatively to send an electron beam into the plasma. Such a beam excites plasma waves near ω_p. These high-frequency oscillations are picked up by a probe and observed on a spectrum analyzer. If a peak in the response can be detected (Fig. 35), it will likely be near $\omega = \omega_p$, and this gives the plasma density. Various spurious effects, such as multiple peaks or surface waves, can cause the resonant ω to differ from ω_p, but when the signal is clear, a good estimate of n can be obtained.

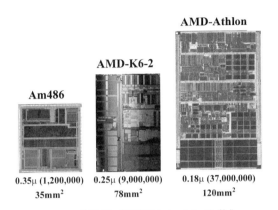

Am486
0.35μ (1,200,000)
35mm²

AMD-K6-2
0.25μ (9,000,000)
78mm²

AMD-Athlon
0.18μ (37,000,000)
120mm²

Courtesy of AMD (SDC Director John Caffal)

Fig. 1. The evoluation of microprocessors shown with the device critical dimension and the packing density.

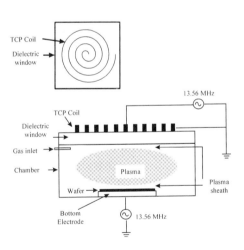

Fig. 2. A schematic diagram of a high density transformer coupled plasma (TCP) reactor.

PRINCIPLES OF PLASMA PROCESSING
Course Notes: Prof. J. P. Chang

PART B1: OVERVIEW OF PLASMA PROCESSING IN MICROELECTRONICS FABRICATION

I. PLASMA PROCESSING

Plasma processing is the most widely used chemical process in microelectronic industry for thin film deposition and etching. Its application extends to surface cleaning and modification, flat panel display fabrication, plasma spary, plasma microdischarge and many other rapidly growing areas. The fundamental understanding of plasma processes is now sufficient that plasma models are emerging as tools for developing new plasma equipment and process, as well as diagnosing process difficulties. In addition, plasma diagnostics are now being implemented as process monitors, endpoint detectors, and process controllers to improve processing flexibility and reliability.

Take plasma etching processes for example, high density plasma reactors have been developed to address the challenges in patterning features less than 0.25 μm with high aspect ratios (Fig. 1). The challenges include maintaining etching uniformity, etching selectivity, high etching rate, and reducing the substrate damage. Various high density plasma sources such as transformer coupled plasma (TCP) and electron cyclotron resonance (ECR) reactors have been developed to achieve high fidelity pattern transfer for manufacturing of ultra large scale integrated (ULSI) electronic devices (Fig. 2).

In a transformer coupled plasma (TCP) reactor, a spiral planar inductive coil is mounted on a dielectric window on the reactor. Plasma is generated by coupling the oscillating radio frequency (rf) magnetic field (13.56 MHz) inductively. Plasma sheath, a dark space between plasma and the electrodes, is developed due to the different mobility of electrons and ions. The bottom electrode can be powered by a separate rf source to control the ion bombardment energy. The energetic ions and reactive neutrals produced are highly reactive, thereby facilitating surface (and/or gas phase) reactions with lower activation energies, and enhance greatly the reaction kinetics.

In plasma processes for the fabrication of microelectronics, DC or rf glow discharges are used

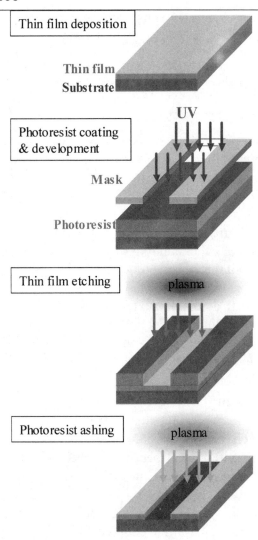

Fig. 3. Photolithography process flow.

Table 1. Typical operating ranges of a glow discharge used in microelectronic fabrication.

Property	Range
Pressure	0.001-10 torr
Electron density	
Low density	10^9-10^{11} cm^{-3}
High density	10^{11}-10^{13} cm^{-3}
Average electron energy	1-10 eV
Average neutral/ion energy	0.025-0.05 eV
Ionized fraction of gas	
Low density	10^{-7}-10^{-5}
High density	10^{-3}-10^{-1}
Neutral diffusivity	20-20,000 cm^2/s
Free radical density	5-90%
Power dissipation	0.1-10 W/cm^2

to etch, deposit, sputter, or otherwise alter the wafer surfaces. These plasmas produce highly reactive neutrals and ions at low temperatures by the introduction of energy into the plasma through its free electrons that in turn collide with the neutral gas molecules.

II. APPLICATIONS IN MICROELECTRONICS

Plasmas are used in several major microelectronics processes: sputtering, plasma-enhanced chemical vapor deposition (PECVD), plasma etching, ashing, implantation, and surface cleaning/modification, each is described below and a few process steps are shown schematically in Fig. 3:

Deposition:
- Semiconductor (silicon)
- Metal (aluminum, copper, alloys)
- Dielectric (silicon dioxide, silicon nitride, metal oxides, low-k dielectrics)

Etching:
- Semiconductor (silicon)
- Metal (aluminum, copper, alloys)
- Dielectric (silicon dioxide, silicon nitride, metal oxides, low-k dielectrics)

Ashing:
- Photoresist removal
- Photoresist trimming

Implantation:
- Dopant ion implant (B$^+$, P$^+$, As$^+$ ...etc.)

Surface Cleaning / Modification:
- Contamination removal
- Modification of surface termination

In each the plasma is used as a source of ions and/or reactive neutrals, and is sustained in a reactor so as to control the flux of neutrals and ions to a surface. The typical ranges of properties for a glow discharge used in microelectronic fabrication are as shown in Table 1.

In sputtering processes (Fig. 4), ions are extracted from a plasma, accelerated by an electric field, and impinge upon a target electrode composed of the material to be deposited. The bombarding ions dissipate their energy by sputtering processes in which the surface atoms are ejected primarily by momentum transfer in collision cascades. The ejected atoms are deposited upon wafers that are placed within line-of-sight of the target electrode, thus facilitating the vapor transport of material

$$Ta_{solid} + Ar^+ \rightarrow Ta_{gas} \rightarrow Ta_{film}$$

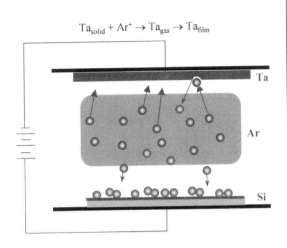

Fig. 4. Sputtering deposition process

$$Si(OC_2H_5)_4 + e^- \rightarrow Si(OC_2H_5)_3(OH) + C_2H_4 + e^-$$

$$O_2 + e^- \rightarrow 2O + e^-$$

$$O + Si(OC_2H_5)_3(OH) \rightarrow Si(OC_2H_5)_2(OH)_2 + C_2H_4O$$

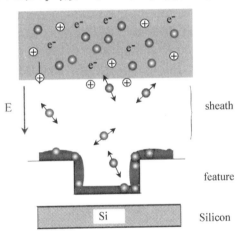

Fig. 5. A plasma-enhanced chemical vapor deposition process.

without appreciably heating either the target electrode or the wafer on which the film is deposited.

Plasma enhanced chemical vapor deposition uses a discharge to reduce the temperature at which films can be deposited from gaseous reactants through the creation of free radicals and other excited species that react at lower temperatures within the gas-phase and on the surface (Fig. 5). The quality of the deposited film often can be improved by the use of the plasma ion flux to clean the surface before the deposition begins and by heating during processing. In addition, the ion flux can alter the film during deposition by cleaning, enhancing the mobility of adsorbed species, etc.

In plasma etching, as shown in Fig. 6, the plasma produces both highly reactive neutrals (e.g., atomic chlorine) and ions that bombard the surface being etched. The neutrals react with the surface to produce volatile species that desorb and are pumped away. Ion bombardment often increases the etching rate by removing surface contaminants that block the etching or by directly enhancing the kinetics of the etching. Ultra large scale integration (ULSI) requires the etching of films with thickness comparable to their lateral feature dimensions. Directional plasma etching processes must be used to pattern such features to obtain the necessary fidelity of pattern transfer. Wet etching processes (which use aqueous acids or bases) and chemical based plasma etching processes are typically isotropic, and produce undercutting of the pattern at least equal to the film thickness.

An ideal plasma etching process requires perfect pattern transfer by anisotropic (directional) etching of polysilicon (the portion not protected by the photoresist), and no etching of either photoresist or silicon dioxide upon ion bombardment (infinite selectivity). This typically requires highly directional ions and minimal spontaneous etching of polysilicon by reactive neutrals.

In reality, many non-ideal factors including transport of reactive species into the feature and interactions of reactive species with the surface affect etched profiles. For example, ions undergo collisions across the sheath, bear a finite angular distribution, and affect the etching anisotropy. As the aspect ratio (depth/width) of the feature

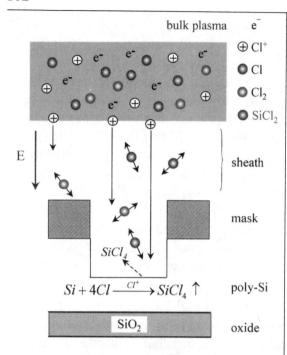

Fig. 6. Chlorine ion-enhanced etching of photoresist patterned polysilicon. The major reactive species in plasma include energetic chlorine ions (Cl^+) and reactive neutrals (Cl, Cl_2, $SiCl_2$).

increases, shadowing effect of the neutral species due to their non-zero reaction probabilities on the sidewall of the feature can cause concentration gradients within the feature and significantly alter the etching profiles and the etching uniformity. The etching products or by-products with high sticking probabilities can deposit on the surfaces within the feature and alter the profile evolution. Specifically, etching of photoresists or electron impact dissociation of the etching products ($SiCl_4$) lead to the formation of carbonaceous contamination and $SiCl_2$ that form passivation layers in the feature and prevent the sidewall from being etched. The balance between simultaneous etching and deposition controls the overall profile topography change during plasma etching processes. Moreover, shadowing of the isotropic electrons and positive charging on the silicon dioxide surface in etching high aspect ratio (width/depth) features can build up an electrical field on the oxide surface to distort the ion trajectory. These etching phenomena are highly convoluted and a thorough understanding of the fundamental mechanisms by which the etching anisotropy is achieved is required to develop rapid, directional, high resolution and damage-free etching processes.

Photoresist ashing is typically done with an oxygen plasma, as shown in Fig. 7. Resist trimming allows finer feature definition and will be detailed in the Epilogue. Other processes including ion implantation, surface modification, and surface preparation will be discussed in the later part of this course.

Major References for Part B

1. H. H. Sawin, Plasma processing for microelectronic fabrication, Lecture notes, MIT (1996).

2. B. Chapman, Glow discharge processes, Wiley (1980).

3. M. A. Lieberman and A. J. Lichtenberg, Principles of plasma discharges and materials processing, Wiley (1994).

4. J. R. Hollahan and A. T. Bell, Fundamentals of plasma chemistry: Techniques and applications of plasma chemistry, Wiley (1974).

5. J. F. O'Hanlon, A User's guide to vacuum technology, Wiley (1989).

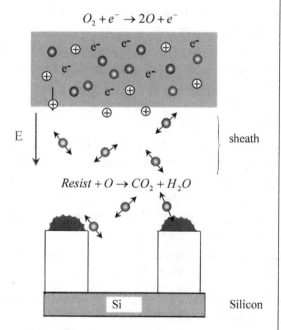

Fig. 7. An oxygen plasma ashing photoresist process.

PRINCIPLES OF PLASMA PROCESSING
Course Notes: Prof. J. P. Chang

PART B2: KINETIC THEORY AND COLLISIONS

I. KINETIC THEORY

The particle velocity distribution is critical in plasma science, and can be derived from the *Boltzmann transport equation*. The transport properties of the electrons and the ions can be derived once their velocity distributions are determined. As described below, the electron-velocity distribution is a function of the strength of the applied electric field that sustains the plasma, and the collisions of neutral and charged species within the plasma. The electron-velocity distribution function, $f_e(\mathbf{r}, \mathbf{v}, t)$, represents the concentration of electrons at a particular position in space, $\mathbf{r}$, that have a velocity vector, $\mathbf{v}$, at a given time, t (Fig. 1). Thus, the number of electrons at a position $\mathbf{r}$ and t is

$$n(\mathbf{r}, t) = \int_{-\infty}^{\infty} f_e d\mathbf{v} \qquad (1)$$

where $d\mathbf{v}$ represents a triple integration over the three coordinates of velocity space. The average value of a property, $\phi(\mathbf{r}, \mathbf{v})$, of the electrons can be calculated as

$$n < \phi > = \int_{-\infty}^{\infty} \phi f_e d\mathbf{v} \qquad (2)$$

For example, the mean velocity $\mathbf{v}(\mathbf{r}, t)$ is:

$$<v> = \bar{v} = \frac{1}{n} \int_{-\infty}^{\infty} \mathbf{v} f_e d\mathbf{v} \qquad (3)$$

and the particle flux is:

$$\Gamma(\mathbf{r}, t) = n\mathbf{v} = \int_{-\infty}^{\infty} \mathbf{v} f_e d\mathbf{v} \qquad (4)$$

We will learn how to determine these terms very soon.

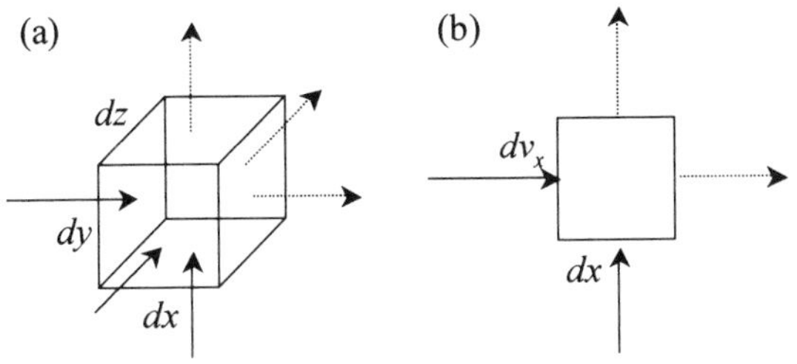

The Boltzmann transport equation is an electron balance over a differential volume of the real and velocity spaces, defined by $d\mathbf{r} d\mathbf{v}$:

$$\frac{\partial f_e}{\partial t} + \mathbf{v} \cdot \nabla_r f_e + \mathbf{a} \cdot \nabla_v f_e = \left(\frac{\partial f_e}{\partial t} \right)_c \qquad (5)$$

The first term represents the unsteady-state fluctuations in the element, i.e., the change in the number

Fig. 1. Differential volume for electron balance: (a) three-dimensional x-y-z phase space, (b) one-dimensional x-v_x phase space. The solid arrows indicate particles entering the differential volume, while the dotted arrows indicate particles leaving the differential volume.

of electrons within the volume that have a certain velocity. The second term represents the convective flux of electrons in space due to the net number of electrons lost or gained by movement of electrons to and from the volume. The third term is the convective flux of electrons from the volume caused by their coupling to the electric field, due to their acceleration by the electric field. If there is only an electric field, we can write $\mathbf{a}$ as $\dfrac{e\mathbf{E}}{m}$. The right-hand-side term accounts for the transfer of electrons to and from the differential volume by collisions. The collisions can be quantified as:

$$\left(\frac{\partial f_e}{\partial t}\right)_c = \int d\mathbf{v}\int_0^{\pi}\left(f_e' f_T' - f_e f_T\right)v_R\sigma_c 2\pi\sin\theta d\theta \quad (6)$$

where electron from distribution f_e has a velocity v and a target particle T from distribution f_T has a velocity of v_T. After the collision, the particles are scattered to the primed velocities in the primed distributions. Note that v_R is the relative velocity ($=|v-v_T|$), σ_c is the collision cross-section, and θ is the scattering angle. The product $f_e f_T$ is associated with the particles before collision and $f_e' f_T'$ are the distributions after a collision has occurred. The collision event, thus, can either remove electrons from a given differential volume or promote them by the reverse process.

To solve the Boltzman equation and obtain the electron energy distribution function (EEDF), the electron distribution can be broken into isotropic and anisotropic terms, as shown in equation (7). The isotropic term f_e^i represents all the electrons that have random direction and a velocity distribution that is invariant with direction. The anisotropic term f_e^a represents those electrons that have a favored direction. It is largest when $\mathbf{v}$ is in the direction of the gradient and smallest when $\mathbf{v}$ is perpendicular to the gradient.

$$f_e = f_e^i + \frac{\mathbf{v}}{v}\cdot \boldsymbol{f}_e^a \quad (7)$$

Assuming that there is only an electric field, we can substitute this expression for f_e into the Boltzmann transport equation and get the following two equations:

$$(8)$$

$$\frac{\partial f_e^i}{\partial t} + \frac{\mathbf{v}}{3} \nabla_r \cdot f_e^a - \frac{e\mathbf{E}}{3mv^2}\frac{\partial}{\partial v}\left(v^2 f_e^a\right) = \left(\frac{\partial f_e^i}{\partial t}\right)_{cc} + \left(\frac{\partial f_e^i}{\partial t}\right)_{elastic} + \left(\frac{\partial f_e^i}{\partial t}\right)_{inelastic}$$

$$\frac{\partial f_e^a}{\partial t} + v\nabla_r f_e^i - \frac{e\mathbf{E}}{m}\frac{\partial f_e^i}{\partial v} = \left(\frac{\partial f_e^a}{\partial t}\right)_{cc} + \left(\frac{\partial f_e^a}{\partial t}\right)_{elastic} + \left(\frac{\partial f_e^a}{\partial t}\right)_{inelastic} \quad (9)$$

Here the collisional term has been expanded to the right-hand-side terms that represent the coulombic, elastic, and inelastic collision loss terms. The coulombic interaction term is significant only when the fraction of ionization within the plasma is greater than 10^{-5}. The elastic collisions between the electrons and the neutrals lead to a general heating of the neutrals, but result in little energy exchange due to the large difference in particle masses. They account for a significant randomization of electron angular velocity distribution, and therefore greatly reduce the anisotropic nature of the electron velocity distribution. The inelastic collisions account for very little of the transport properties of the electrons. They do, however, have a major effect on the distribution of the electron velocities, and on the physical and chemical properties of the plasma. It is the inelastic collisions that are responsible for the generation of excited species and ions. The detailed forms of these collisional terms can be found in the reference by Hollahan and Bell.

The solutions to these two equations are complex, therefore proper assumptions are often used to obtain much simpler solutions for analysis. First, we assume that an alternating electric field, $E_o e^{-i\omega t}$, is imposed on a homogeneous plasma. From equation (9), we further assume that the coulombic and inelastic collisions are not important, the elastic collision term is $-\nu_m f_e^a$, and the time dependence of f_e^a is given by $e^{-i\omega t}$. Note that $\nu_m = Nv\sigma_m(v)$ is the momentum transfer frequency for electrons and ω is the frequency of the field. The anisotropic term in the distribution function can then be derived to be:

$$f_e^a = \frac{e\mathbf{E}}{m(\nu_m - i\omega)}\frac{\partial f_e^i}{\partial v} \quad (10)$$

Introduce equation (10) into (8), assume that the isotropic term may be considered time invariant for frequencies greater than the relaxation time of the electron energy by elastic collisions, and neglect the spatial variation, the isotropic term of the distribution function can thus be calculated, and is found to be of the same form as that under a dc field with an effective amplitude E_e given by

$$E_e = E_o \sqrt{\frac{v_m^2}{2\left(v_m^2 + \omega^2\right)}} \qquad (11)$$

Often in calculating the collisional kinetics of a plasma used in microelectronics fabrication, only the isotropic distribution f_e^i is considered. Since the elastic collisions of electrons with the neutrals are very frequent, they sufficiently randomize the field induced directionality of the electrons so that the anisotropic term can be neglected.

Case (a): If the sinusoidal amplitude E_o is 0, and inelastic collisions can be neglected, the isotropic term of the Boltzmann distribution function reduces to

$$vf_e^i + \frac{kT}{m}\frac{\partial f_e^i}{\partial v} = \left(\frac{\partial f_e^i}{\partial t}\right)\Bigg|_c = 0 \qquad (12)$$

and the solution becomes a *Maxwell-Boltzmann distribution* (MBD):

$$f_e^i = Ce^{-\frac{mv^2}{2kT}} \qquad (13)$$

where the constant C can be determined by the normalization condition. Under these conditions the electrons are in equilibrium with the gas molecules and can be characterized by T, the gas temperature. We will expand our discussion on MBD later.

Case (b): If the electric field strength is low, few electrons will suffer inelastic collisions, since the energy transition for inelastic excitation processes will exceed the energy of most electrons. The Boltzmann equation for this condition after one integration becomes:

$$\frac{\partial f_e^i}{\partial v}\left(\frac{e^2 E_e^2 M}{3m^2 v_m^2} + kT\right) + mvf_e^i = 0 \qquad (14)$$

and its solution becomes the *Margenau distribution*:

$$f_e^i = C\exp\left(-\int_o^v \frac{mvdv}{kT + \frac{e^2 E_e^2 M}{3m^2 v_m^2}}\right) \qquad (15)$$

where the constant C can be determined by the normalization condition.

Case (c): When the elastic collision losses become dominant compared to the thermal energy exchange with the gas, the applied field is sufficiently large so that $\dfrac{e^2 E_e^2 M}{3m^2 v_m^2} \gg kT$, and the field oscillation frequency is much less than that of the collision frequency ($\omega^2 \ll v^2$, i.e., $E_e^2 = \dfrac{1}{2} E_o^2$), the solution of the Margenau distribution can be approximated as:

$$f_e^i = C \exp\left(-\int_o^v \frac{mv\,dv}{\dfrac{e^2 E_o^2 M}{6m^2 v_m^2}}\right) \tag{16}$$

For certain gases such as helium and hydrogen, the elastic collision cross-section varies approximately as the reciprocal of the electron velocity, thus v_m is independent of the electron velocity. Then the Margenau distribution reduces to the Maxwell-Boltznann distribution:

$$f_e^i = C \exp\left(-\frac{\dfrac{mv^2}{2}}{\dfrac{e^2 E_o^2 M}{6m^2 v_m^2}}\right) \tag{17}$$

Comparing equation (17) to (13), we obtain an electron temperature defined as:

$$kT_e = \frac{e^2 E_o^2 M}{6m^2 v_m^2} \tag{18}$$

Since the collision frequency v_m is proportional to the density of gas particles with which the electrons collide, it is proportional to the pressure, *p*. The electron energy, *KT*, is, therefore a function of the electric field to pressure ratio, E_o/p. It should be remembered that in this model, the energy losses due to inelastic collisions are assumed to be less than those due to elastic collisions. This solution is only applicable to very low E_o/p where the electron temperature is low and few inelastic collisions occur.

Case (d): If the collision cross-section is assumed to be independent of the electron velocity, the collision frequency is proportional to the velocity and the Margenau distribution becomes the *Druyvesteyn distribution* of the following form:

$$f_e^i = C \exp\left(-\frac{\left(\frac{mv^2}{2}\right)^2}{\frac{e^2 E_o^2 M}{6mN^2\sigma_m^2}}\right) \qquad (19)$$

Here the collision frequency v_m has been expanded as the product of the neutral density, the velocity, and the collision cross-section $Nv\sigma_m$. Note that the Druyvesteyn distribution varies as e^{-av^4}, and again the dependence upon E_o/p is obvious.

In comparison, in the Maxwellian and the Margenau distributions, the high energy tail decreases as the exponential to the negative second power of electron velocity, while the Druyvesteyn distribution declines as the exponential of the negative fourth power. Thus, the Druyvesteyn distribution predicts fewer high energy electrons that can produce ions upon collision, as shown in Fig. 2.

In both the Margenau and the Druyvesteyn distributions, the exponent's argument contains the ratio of the effective electric field to the pressure of the neutrals within the plasma, the ratio E_o/p can be used to characterize the electron distribution function. Thus the average electron energy is also a function of E_o/p. For example, if the pressure of the plasma is increased by a factor of two, a similar electron distribution function can be maintained by increasing the electric field by two.

As stated earlier, the Maxwellian and Druyvesteyn distributions are reasonable solutions for the electron energy distributions for low E_o/p where the inelastic collisions can be neglected, however, these conditions are not valid for steady-state discharges used in processing. To sustain the discharge, sufficient ionization (inelastic collision) must occur to balance the loss to surfaces. Numerical methods have been used to compute the electron energy distribution functions considering inelastic collision as random two-body coulombic collisions. Shown in Fig. 3 is the energy distributions for a hydrogen plasma computed by numerically solving the Boltzmann equation for different extents of ioniza-

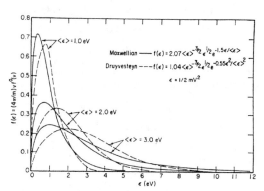

Fig. 2. Comparison of Maxwellian and Druyvesteyn energy distribution functions

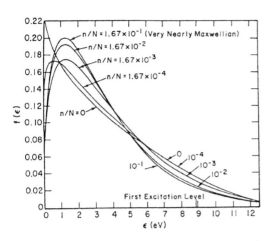

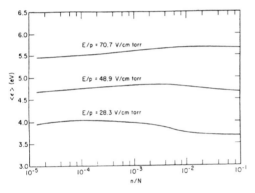

Fig. 3: Effect of the extent of ionization on the energy distribution function for a hydrogen plasma; E/p = 28.3 V/cm-torr (H. Dreicer).

Fig. 4: Variation of average electron energy with the extent of ionization for a hydrogen discharge (H. Dreicer).

tion, n/N, as a constant E_o/p=28.3 V/cm-torr. As the ionization increases, the distribution becomes more Maxwellian-like. This is a result of the increased electron-electron scattering making the energy exchange within the electron population rapid and thereby forming an electron gas like distribution and reducing the importance of the accelerating fields on the electrons.

The effect of electron interactions on the average energy of the plasma is shown in Fig. 4. The electron temperature is quite invariant with the power (i.e., plasma density) for plasmas of sufficient densities, but is largely a function of E_o/p.

In conclusion, due to inelastic collisions, the energy typically is not proportional to the square of the E_o/p as suggested by Margenau and Druyvesteyn equations, but increases with a power varying between 0.5 to 1.

II. PRACTICAL GAS KINETIC MODELS AND MACROSCOPIC PROPERTIES

A low pressure gas ($\leq$ 1 atmosphere) can be modeled using Maxwell-Boltzmann kinetics with deviations of less than a few percent from experimental data for most properties. The gas model assumes that:

a) The gas is composed of rigid spheres that have vanishingly small volumes.

b) There are no long range interactions between spheres.

c) All gas-gas collisions are elastic.

d) The spheres have no rotational or vibrational energy.

The hard sphere model utilizes the concept of a collision diameter, d_o. The collision diameter can be thought of as the distance between the centers of two hard spheres during a collision, as shown in Fig. 5; d_o for a molecule is typically about 3 Å. The collision diameter of two dissimilar molecules is the mean of the collision diameters of the molecules. In reality, there are interactions between molecules for distances greater than their collision diameter; however, these forces are relatively short range and can be neglected if the gas has a low number density (i.e., at low pressures).

Note here that we can also derive several particle conservation equations from Boltzmann trans-

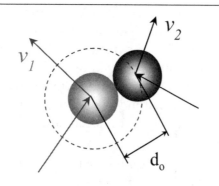

Fig. 5. Illustration of molecular interactions.

port equation, including

Continuity equation:

$$\nabla \cdot (n\mathbf{v}) = \nu_{ion} n_e \qquad (20)$$

Momentum conservation equation:

$$mn\frac{\partial \mathbf{v}}{\partial t} = qn\mathbf{E} - \nabla p - mn\nu_m \mathbf{v} \qquad (21)$$

Energy conservation equation:

$$\nabla \cdot \left(\frac{3}{2} p\mathbf{v} \right) = \left. \frac{\partial}{\partial t} \left(\frac{3}{2} p \right) \right|_c \qquad (22)$$

Boltzmann relation:

$$n_e(\mathbf{r}) = n_o e^{\frac{eV(\mathbf{r})}{KT_e}} \qquad (23)$$

1. Maxwell-Boltzmann Distribution (MBD)

The number of gas particles with a given speed is related to the energy content of the gas, and to the rate at which they collide with each other and with surfaces. A quantitative knowledge of this distribution is needed to calculate properties of the gas and chemical reaction rates. Classical thermodynamics for the kinetic gas model outlined in the previous section dictates a Maxwell-Boltzmann distribution function for the translational speeds of the gas particles. This function can be obtained by a maximization of the entropy (disorder) of a system that contains a given amount of energy. This is equivalent to saying that the system relaxes to a state of maximum disorder for a given quantity of energy. For this probability distribution, the normalized number of particles per unit volume with speed v is

$$P(v) = 4\pi \left[\frac{M}{2\pi KT} \right]^{3/2} v^2 \exp\left[-\frac{Mv^2}{2KT} \right] \qquad (24)$$

where K is Boltzmann's constant, T is the temperature of the gas, and M is the mass of the gas particles. A plot of the Maxwell Boltzmann distribution is shown in Fig. 6. This distribution is sometimes referred to as the Boltzmann distribution or the system can be said to be Maxwellian. The most probable (mp), mean (m), and root-mean-square (rms) velocities are shown in Fig. 6:

$$v_{mp} = \sqrt{\frac{2KT}{M}} \qquad (25)$$

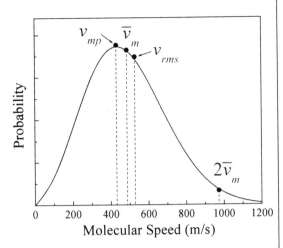

Fig. 6. A schematic of the MBD function

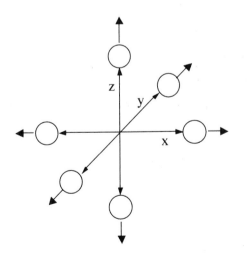

Fig. 7. Cartesian coordinates for illustrating gas particle motions.

$$v_m = \bar{v} = \int_0^\infty v P(v)\,dv = \sqrt{\frac{8KT}{\pi M}} \tag{26}$$

$$v_{rms} = \left(\int_0^\infty v^2 P(v)\,dv\right)^{\frac{1}{2}} = \sqrt{\frac{3KT}{M}} \tag{27}$$

2. A Simplified Gas Model (SGM)

This model's assumptions are the same as those described above, but rather than using the Maxwell-Boltzmann distribution function, a simplified function is used. The simplified distribution assumes that all gas particles have identical speeds where

$$\bar{v} = \sqrt{\frac{8KT}{\pi M}} \tag{28}$$

It is also assumed that one sixth of the gas particles travel along each of the Cartesian axes, as shown in Fig. 7.

For argon at 20°C, the average thermal velocity is 4×10^4 cm/s. It should be noted that by using the Maxwell-Boltzmann distribution in terms of the speed rather than velocity, we have avoided the integration of a vector quantity over a three dimensional velocity space; this allows us to quickly estimate the plasma properties though the results differ slightly to that obtained with MBD (note the results are within the same order of magnitude).

We will use both MBD and SGM to determine some basic properties of the plasma and compare their differences.

3. Energy Content

The energy content of the gas in both models is assumed to be entirely associated with the translational energy of the gas particles (i.e., we neglected rotational, vibrational, and electronic energy). In SGM, since the particles are all going at the same speed, they have equal energies:

$$\text{Energy} = \frac{\overline{mv^2}}{2} = \frac{4KT}{\pi} \tag{29}$$

Using the MBD, we get

$$\frac{\overline{mv^2}}{2} = \frac{\int_0^\infty mv^2 P(v)\,dv}{2} = \frac{3KT}{2} \tag{30}$$

4. Collision Rate Between Molecules

The collision rate between molecules determines the rate at which energy can be transferred within a gas, and the rate at which chemical reactions can take place. To derive the frequency of collisions between gas molecules in SGM, consider a molecule going in the +z direction. In the simplified model, it will have no collision with any other molecules going in the +z direction, since they are all going at the same speed. It will have a closing speed of 2 relative to that of the n/6 molecules going in the -z direction, and a speed of $\sqrt{2}v$ relative to the 4n/6 molecules going in the ±x and ±y directions. Therefore, during a given period of time, the molecule sweeps out a cylindrical volume using d_o as the diameter and with a length equal to the relative speed multiplied by the time. The molecule will suffer a collision if its center comes within a distance d_o of another molecule's center (Fig. 8). The number of collisions per unit time is thus, the swept volume multiplied by the number of molecules per unit volume:

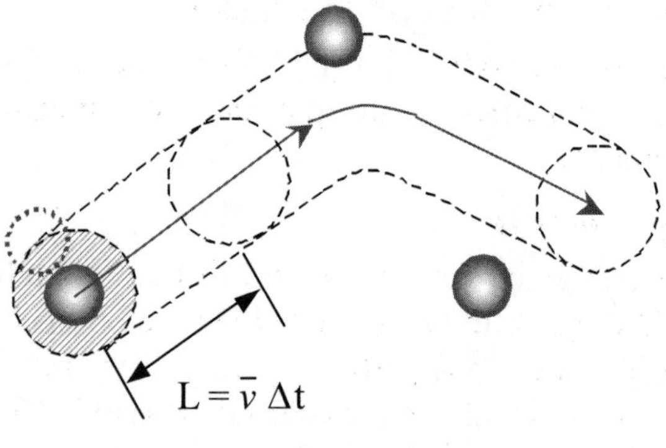

The number of collision is:

$$v = \frac{n}{6}\pi d_o^2 2\bar{v} + \frac{4n}{6}\pi d_o^2 \sqrt{2}\,\bar{v}$$

$$= \xi n\pi d_o^2 \bar{v} = \xi n\pi d_o^2 \sqrt{\frac{8KT}{\pi m}} = \xi p d_o^2 \sqrt{\frac{8\pi}{mKT}} \qquad (31)$$

where $\qquad \xi = \frac{1}{3} + \frac{2}{3}\sqrt{2} = 1.276$

$$p = nKT \text{ (ideal gas law)} \qquad (32)$$

Here the ideal gas law has been used to formulate the number of collisions as function of pressure.

Using the MBD, the constant ξ is evaluated as $\sqrt{2}$, yielding:

$$v = 4pd_o^2 \sqrt{\frac{\pi}{mKT}} \qquad (33)$$

For argon at 20°C and 1 torr of pressure, the collision frequency is $6.7 \times 10^6 \text{ s}^{-1}$.

5. Mean Free Path

The mean free path, λ, is the average distance traveled by a gas molecule between collisions. It is a determining factor in the rate at which a gas particle can diffuse within a gas. For a given time increment, a molecule travels a distance equal to its

$$L = \bar{v}\,\Delta t$$

Fig. 8. The volume swept by a molecule equals $\pi d_o^2 (speed)\Delta t$

speed multiplied by the elapsed time. During this time it suffers the number of collisions determined by the previous derivation of the collision frequency. Thus, the mean free path, λ, in SGM is

$$\lambda = \frac{\bar{v}\,\Delta t}{v\Delta t} = \frac{\bar{v}}{v} = \frac{1}{\xi n \sigma_c} = \frac{KT}{\xi p \pi d_o^2} \qquad (34)$$

Using the MBD, a very similar form is obtained:

$$\lambda = \frac{KT}{\sqrt{2}\, p \pi d_o^2} \qquad (35)$$

For argon at 20°C and 1 torr, the mean free path is 6×10^{-3} cm or 60 microns.

To calculate the probability, $P(x)$, that a molecule will travel a distance x without suffering a collision, we can relate the change in the probability, $dP(x)$, to the probability that it will suffer a collision in the next increment of distance, dx. Thus, the probability that a collision will occur in the distance between x and $x + dx$ is the product of the probability that it will have reached x without a collision, the number of collisions per unit length, and the incremental distance dx:

$$dP(x) = -P(x)\frac{dx}{\lambda} \qquad (36)$$

Integrating the above equation, we find that the probability of not suffering a collision decays exponentially with increasing x:

$$P(x) = \exp\left(-\frac{x}{\lambda}\right) \qquad (37)$$

Since the mean free path is a function of the ratio of average gas molecule velocity to the collision cross-section of the molecule,

$$\lambda = \frac{\bar{v}}{v} = \frac{1}{\xi n \sigma_c} \qquad (38)$$

substituting for λ in the probability expression yields

$$P(x) = \exp\left(-\xi n \sigma_x x\right) \qquad (39)$$

For example, the probability that a molecule travels one mean free path without suffering a collision is the exp(-1), or 37%, while the probability that a molecule travels three mean free path without suffering a collision is the exp(-3), or 5%.

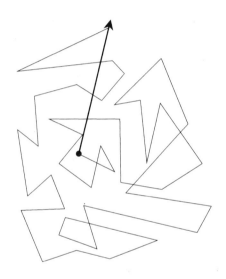

Fig. 9. Random motion and mean free path.

6. Flux of Gas Particles on a Surface

The flux of gas molecules striking a surface determines the maximum rate at which the gas can react with the surface, assuming unity sticking probability -- every particle that strikes sticks to the surface. The term flux refers to the number of molecules that impinge on a unit surface area in a unit time increment (Fig. 10). Since all the molecules have speed $\bar{v}$ in SGM, the "effective" volume of gas that can strike the surface within time t is

$$\text{Volume} = \bar{v}\,t \text{ (unit area)}$$

This is an effective volume because only a fraction ($n/6$) of the molecules in the gas are traveling in the correct direction, i.e., towards the surface. The number of molecules striking a unit area in time t is the effective volume multiplied by $n/6$, so that the flux on the surface is

$$\text{Flux} = \frac{n\bar{v}}{6} = \frac{n}{6}\sqrt{\frac{8KT}{\pi m}} \qquad (40)$$

Using the MBD, the flux is calculated by integrating over the velocity component perpendicular to the surface, yielding

$$\text{Flux} = \frac{n\bar{v}}{4} \qquad (41)$$

Note that the Boltzmann distribution gives a factor of 1/4 rather than 1/6. For argon at 20°C and 1 torr, the number flux is 3×10^{20} molecules/cm^2. To put this number in perspective, a monolayer contains approximately 10^{15} molecules/cm^2, so that about 1 monolayer equivalent molecules strike the surface per second at 10^{-6} torr. If the sticking coefficient of these molecules is unity, one monolayer of material will be deposit at 10^{-6} torr every second!

7. Gas Pressure

The pressure exerted by a gas on a surface can be calculated from the change of momentum that the gas molecules undergo during collisions with the surface. For a system in thermodynamic equilibrium, the gas leaving the surface must have the same speed (and velocity) distribution as that strikes the surface. Pressure is the force per unit area or the change in the gas momentum per unit area per unit time, as shown in Fig. 11:

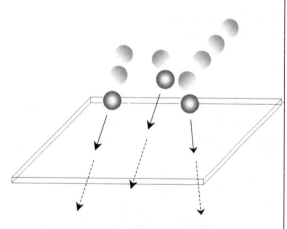

Fig. 10. Fluxes to a surface.

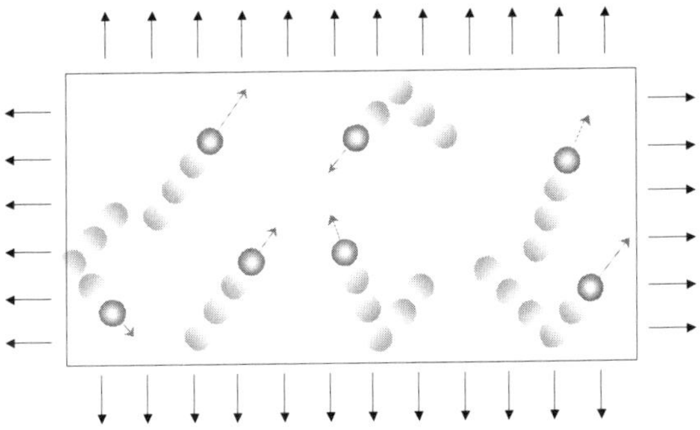

$$\text{Pressure} = \frac{\text{Force}}{\text{Area}} = \frac{(\text{Momentum change/molecule})\text{Flux}}{\text{Area}}$$

In SGM, each molecule striking the surface undergoes an average momentum change of $2m\bar{v}$. The pressure, p, for the simplified model is then the product of the number flux calculated above and the momentum change per collision,

$$p = 2m\bar{v}\frac{n\bar{v}}{6} \qquad (42)$$

substituting for $\bar{v}$ gives

$$p = \frac{8}{3\pi}nKT \qquad (43)$$

Using the MBD with proper integration:

$$p = \int_0^\infty 2mvP(v)dv\Gamma = nKT \qquad (44)$$

which is the ideal gas law. Using the ideal gas law, the number density, n, of argon at 20°C and 1 torr is y a fact^{or} of two, a si^{m}ilar electron distribution van der Waals equation of state takes into account the repulsive forces at work at short distances which tends to reduce V and the attractive forces at longer distances that tends to decrease p:

$$p = \frac{nRT}{V-nb} - \frac{n^2 a}{V^2} \qquad (a, b \text{ are constants})$$

$$\left(p + a\frac{n^2}{V^2}\right)(V - nb) = nRT \qquad (45)$$

There are a number of different pressure units used in vacuum calculations, and the following conversion relations contain the commonly used units:

1 atmosphere = 760 mmHg = 760 torr = 10^5 Pascal

1 bar = 10^5 Pascals

1 mtorr = 1×10^{-3} torr = 1 μm Hg = 1 micron

1 Pascal = 9.8 kg/m-s^2 = 1 N/m^2 = 7.6 mtorr

8. Transport Properties

The flux of heat, mass, and momentum are considered the transport properties of a medium. These fluxes are generally calculated using the thermal conductivity k_T, the mass diffusivity D, and

Fig. 11. Pressure due to the impulse times the rate of collision with the wall.

the viscosity η. The form of the flux equations is shown below.

Heat Flux = $q_z = -k_T \dfrac{dT}{dz}$ (temperature gradient)

Mass Flux = $j_z = -D \dfrac{dn_i M_i}{dz}$ (conc. gradient)

Momentum Flux = $p_z = \eta \dfrac{dv_y}{dz}$ (velocity gradient)

These coefficients can be calculated using the MBD, yielding:

$k_T = \dfrac{n c_v \bar{v} \lambda}{3}$ $\rightarrow$ proportional to $T^{1/2}$ $M^{-1/2}$ p^o $\left(\pi d_o^2\right)^{-1}$

$D = \dfrac{\bar{v} \lambda}{3}$ $\rightarrow$ proportional to $T^{3/2}$ $M^{-1/2}$ p^{-1} $\left(\pi d_o^2\right)^{-1}$

$\eta = \dfrac{n M \bar{v} \lambda}{3}$ $\rightarrow$ proportional to $T^{1/2}$ $M^{1/2}$ p^o $\left(\pi d_o^2\right)^{-1}$

where c_v is the heat capacity at constant volume.

Typical values of gas diffusivity are 0.2 and 150 cm^2/s at 1 atm and 1 torr, respectively. The diffusion of mixtures requires appropriate mixing rules. It should be noted that mass diffusivity is inversely proportional to pressure, so that the diffusivities at plasma processing pressures are very large (100-10,000 cm^2/s).

9. Gas Flow

The flow of gas must be considered when designing a pumping system for a plasma reactor. Pumping low pressure gases is surprisingly difficult, and large conduits are required to provide the necessary conductivity.

The flow of gases in a conduit or chamber can be divided into three regimes, based on the ratio of the gas mean-free-path to the inner dimension of the enclosure (the Knudsen number): $Kn = \lambda/L$, as:

a) Viscous flow or continuum regime: $\lambda/L < 0.01$

b) Transition regime: $0.01 < \lambda/L < 1$

c) Molecular flow or rarefied regime: $\lambda/L > 1$

Table 1 summaries the mean free path of gas molecules at various pressures and the corresponding flow regimes. In the viscous regime the collisions between gas molecules dominate in determining the flow characteristics. A no-slip boundary condition exists at any surface, meaning that the velocity of the gas is zero at the surface. For most situations in vacuum systems, the flow is laminar, and has flow properties similar to that of molasses. This is an artifact of the independence of gas viscosity with respect to pressure. The fluid flow characteristics are dependent upon the Reynolds number which is $\rho v D / \eta$, where ρ is the density per unit volume. A Reynolds number of less than 2000 in a circular conduit indicates laminar flow. At higher pressures and high velocities, the flow can become turbulent.

In the molecular flow regime, any sort of continuum model for the flow behavior of the gas breaks down; the majority of the gas collisions occur with the walls of the conduit, not between gas molecules. In this regime the pressure does not influence the flow characteristics, as shown in Fig. 12. These flow regimes as a function of the gas density are summarized in Fig. 13.

The flow characteristics in the transition regime are not fully understood. This makes the analysis of plasma processing challenging since most of the plasma processes used in microelectronics processing are performed in this regime.

The flow of a gas through a conduit is usually characterized as a conductance, analogous to the conductance in an electrical circuit. The conductance, C, is the ratio of the flow rate in units of pressure-volume/time (Q) and the pressure difference:

$$C = \frac{Q}{p_1 - p_2} \qquad (46)$$

Note that the conductance has units of volume/time. As in the electronic circuits, the overall conductivity of a system can be calculated using

$$\frac{1}{C_{\text{overall}}} = \frac{1}{C_1} + \frac{1}{C_2} + \frac{1}{C_3} + \ldots \qquad (47)$$

Use a circular tube with a length much greater than its diameter as an example, the conductivity for air in it in the rarefied regime is given by

$$C_{\text{long tube}} = 12.2 \frac{D^3}{L} \qquad (48)$$

In the continuum regime, the conductivity is dependent upon the pressure and can be estimated as

Table 1. Mean free path of gas molecules at various pressures.

p(torr)	10^{-4}	10^{-3}	10^{-2}	10^{-1}	1
λ (cm)	50	5	0.5	.05	.005
λ / L	1	0.1	.01	.001	.0001
Flow	R	T	T	V	V

R = Rarefied flow
T = Transition flow
V = Viscous flow

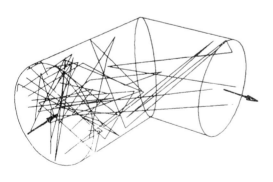

Fig. 12. Computer simulation of 150 particles move in a pipe (molecular flow).

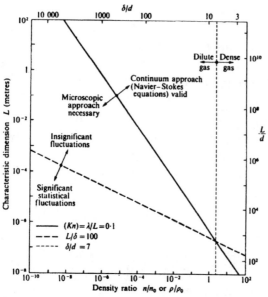

Fig. 13.   Effective limits of major gas flow regimes, $d_o = 3.7 \times 10^{-10}$m. (G. A. Bird, 1976).

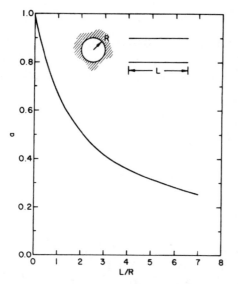

Fig. 14.   Molecular transmission probability, a, of a short tube as a function of the tube length to radius.

$$C_{\text{long tube}} = \frac{\pi D^4}{128 \eta L} \frac{(p_1 + p_2)}{2} \qquad (49)$$

The calculation of flows in the transition regime can be bracketed by calculating for both the rarefied and the viscous flows; i.e., the transition flow is between these two limits.

The conductance of many common conduits encountered in vacuum systems are given in the handbook on vacuum technology by Hanlon. For example, Fig. 14 shows the molecular transmission probability, a, of a short tube as a function of the tube length to radius, and the conductance of a such tube can be calculated as:

$$C_{\text{short tube}} = \frac{a v \pi R^2}{4} \qquad (50)$$

while Fig. 15 shows the molecular transmission probability of a round pipe with entrance and exit apertures. These plots are extremely useful in determining the conductance of a plasma reactor with its pumping elements for design optimization.

The pumping speed of a vacuum pump is typically rated in terms pumping speed, S, where

$$S = \frac{Q}{p} \qquad (51)$$

The pumping speed has units identical to those of conductance; it is sometimes given as a function of pressure, but is often relatively constant over the normal working range (Fig. 16).

The residence time in the system is therefore:

$$\tau = \frac{V}{S} = \frac{pV}{pS} = \frac{pV}{Q} \qquad (52)$$

For example, a gas flow of 5 sccm (standard cubic centimeters per minute) is introduced into a chamber that is pumped by a roughing pump with a pumping speed of 400 l/min through a 2 m long circular tube which is 1 inch in diameter. If the gas flow is assumed to be air, we can calculate the upper limit of the pressure by assuming that the flow is in the rarefied regime to be 0.7 Torr.

With the development of computational fluid dynamic models and software (CFD), the gas velocity and pressure profiles can be calculated quite accurately given the plasma reactor geometry, operat-

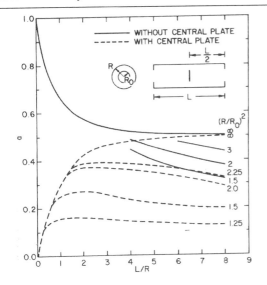

Fig. 15. Molecular transmission probability of a round pipe with entrance and exit apertures.

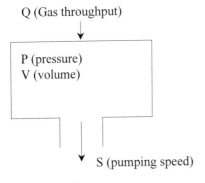

Fig. 16. Schematic of a gas pumping system.

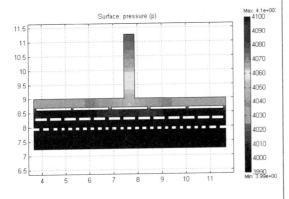

Fig. 17: Schematic of the pressure gradient in a plasma reactor.

ing conditions, and the pumping speed of the system. Fig. 17 shows a schematic of the pressure gradient in a plasma reactor, with one gas injection port, multiple layers of gas distribution rings (showerhead design), and a pumping port.

III. COLLISION DYNAMICS

1. Collision cross sections

The collision cross-section, σ_c, is an averaged value that is proportional to the probability that a certain type of collisional process will occur, and is a function of the closing velocities of the particles. At times it will also be a function of the angle between incident particle trajectory and a line connecting the centers of the particles as they collide; in this case, integrating over all possible collision configurations will yield the correct cross-section.

As shown in Fig. 18, particles incident with impact parameters between h and $h+dh$ will be scattered through angles between θ and $\theta+d\theta$. With central forces, there must be complete symmetry around the central axis so that:

$$2\pi h \cdot dh = -I(v,\theta) \cdot 2\pi \sin\theta \cdot d\theta \qquad (53)$$

The differential cross section, $I(v,\theta)$, is the proportionality constant and is derived as:

$$I(v,\theta) = \frac{h}{\sin\theta}\left|\frac{dh}{d\theta}\right| \qquad (54)$$

therefore the total scattering cross-section, σ_s, is:

$$\sigma_s = 2\pi \int_0^\pi I(v,\theta)\sin\theta d\theta \qquad (55)$$

and the momentum transfer cross-section, σ_m, is:

$$\sigma_m = 2\pi \int_0^\pi (1-\cos\theta)I(v,\theta)\sin\theta d\theta \qquad (56)$$

To solve for the complex scattering trajectories, we can convert the scattering angles from the laboratory frame (Fig. 19) into the center-of-mass (CM) coordinates (Fig. 20), and relate the scattering angles θ_1 and θ_2 to the scattering angle ϕ in the CM frame and the two particles masses: m_1 and m_2:

$$\tan\theta_1 = \frac{\sin\phi}{\cos\phi + \dfrac{m_1}{m_2}} \qquad (57)$$

$$\tan\theta_2 = \frac{\sin\phi}{1-\cos\phi} \qquad (58)$$

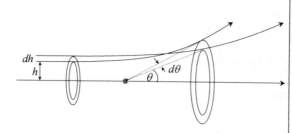

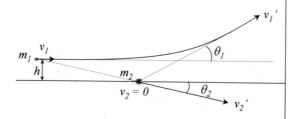

Fig. 18: Schematic illustration of the differential cross-section.

Fig. 19. The relation between the scattering angles in the laboratory frame.

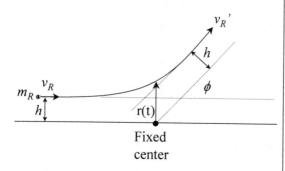

Fig. 20. The relation between the scattering angles in the center of mass (CM) frame.

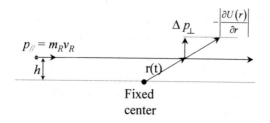

Fig. 21. Small angle scattering with momentum impulse, $\Delta P_\perp$, illustrated.

For electron colliding with ions or neutrals, since $\frac{m_1}{m_2} \ll 1$, the reduced mass $m_R = \frac{m_1 m_2}{m_1 + m_2} \approx m_1 \; (m_e)$, therefore:

$$\theta_1 \approx \phi$$

For equal mass particle collision, $m_1 / m_2 = 1$ and $m_R \approx \frac{1}{2} m_1$, so

$$\theta_1 \approx \frac{\phi}{2}$$

For small angle scattering, we can use the momentum impulse analysis, $\Delta P_\perp$, to solve for the collision cross sections and scattering angles (Fig. 21). Using momentum balance and assume the central force is of a universal form:

$$U(r) = \frac{C}{r^i} \tag{59}$$

Following Smirnov's derivation (1981), we found the scattering angle to be:

$$\phi = \frac{A}{W_R h^i} \tag{60}$$

where $W_R = \frac{1}{2} m_R v_R^2$ is the kinetic energy of the center-of-mass, and the parameter A relates to the Γ function with a constant C:

$$A = \frac{C\sqrt{\pi}}{2} \frac{\Gamma\left(\frac{i+1}{2}\right)}{\Gamma\left(\frac{i+2}{2}\right)} \tag{61}$$

The differential collision cross-section is thus:

$$I(v_R, \phi) = \frac{1}{i} \left(\frac{A}{W_R}\right)^{\frac{2}{i}} \frac{1}{\phi^{2+\frac{2}{i}}} \tag{62}$$

From equation (62), we can then derive the collision cross-sections and collision frequency or reaction rates. For different potential energies, we can determine the dependencies on the relative velocity, as shown in Table 2.

2. Energy Transfer

Consider the simplest scattering of two spheres as representing particle-particle scattering. Choose a frame of reference that is fixed with respect to the more massive sphere, m_2. The lighter

Table 2.
Dependence of collision cross sections and collision frequency (rate constant K) on velocity

	$U(r)$	σ	v or K
Coulombic	$1/r$	$1/v_R^4$	$1/v_R^3$
Perm. dipole	$1/r^2$	$1/v_R^2$	$1/v_R$
Induced dipole	$1/r^4$	$1/v_R$	const.
Hard sphere	$1/r^{i \to \infty}$	const.	v_R

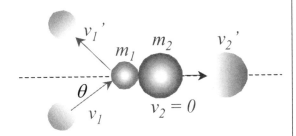

Fig. 22. Illustration of a collision process.

sphere, m_1, is moving with speed v_1 towards m_2. As shown in Fig. 22, θ is the angle between the incident sphere and the line between the centers of the spheres as they collide. After the collision the spheres leave with velocities v_1' and v_2'.

During the collision process both energy and momentum must be conserved. The momentum conservation can be broken into two components, one parallel with the collision centerline and another perpendicular to the centerline and falling in the plane defined by the two lines. Momentum is also conserved in the direction perpendicular to both the centerline and the plane, but since the velocity components in this direction are always zero, it will be ignored.

Parallel Momentum: $m_1 v_1 \cos\theta = m_1 v_{1/\!/}' + m_2 v_2'$

Perpendicular Momentum: $m_1 v_1 \sin\theta = m_1 v_{1\perp}$

Energy: $\frac{1}{2} m_1 v_1^2 = \frac{1}{2} m_1 v_1'^2 + \frac{1}{2} m_1 v_2'^2$

Notice that due to the choice of the centerline as a coordinate, the incident particle does not change its velocity in the perpendicular direction. If E_t is the energy transferred from the incident sphere to the fixed sphere, and E_i is the initial energy of the incident sphere, the fractional energy transferred is given by

$$\zeta = \frac{E_t}{E_i} = \frac{\frac{1}{2} m_2 v_2'^2}{\frac{1}{2} m_1 v_1^2} = \frac{4 m_1 m_2}{(m_1 + m_2)^2} \cos^2 \theta_2 \qquad (63)$$

$$\rightarrow \quad \zeta = \frac{2 m_1 m_2}{(m_1 + m_2)^2} (1 - \cos\phi) \qquad (64)$$

If the masses of the spheres are the same, the fractional energy transferred is

$$m_i = m_t \quad \rightarrow \quad \varsigma = \frac{1}{2}(1 - \cos\phi) \approx \frac{1}{2}$$

at low scattering angles.

This expression would apply for the scattering of ideal-like gas molecules. Note that the fractional energy transfer is unity for head-on collisions in this case. In electron-neutral scattering, the masses are very different and the energy transfer approaches zero in elastic collision:

$$m_i \gg m_t \quad \rightarrow \quad \varsigma \approx \frac{2m}{M} \approx 10^{-4}$$

3. Inelastic Collisions

In the collision process, energy can be transferred from kinetic energy to internal energy if at least one of the particles is complex enough to permit such internal energy states. Atomic gases can undergo electronic transitions where at least one of the outer electrons is promoted into a higher energy state. Diatomic or more complex molecules have rotational and vibrational states that can be excited by inelastic collisions. The energy separation of these states increases in order of rotational, vibrational, and electronic excitations. The possibility of internal energy transfer can be incorporated into the elastic model described above by the inclusion of a ΔU term in the energy conservation relation, where ΔU is the gain in internal energy of the molecule that was struck:

$$\frac{1}{2}m_i v_i^2 = \frac{1}{2}m_i u_i^2 + \frac{1}{2}m_t u_t^2 + \Delta U \qquad (65)$$

The amount of energy transferred upon collision is now not a unique function of the geometry of the collision and the masses of the spheres. It is a continuous function,

$$\Delta U = m_t u_t v_i \cos\theta - \frac{1}{2}\frac{m_t}{m_i}(m_i + m_t)u_t^2 \qquad (66)$$

The maximum energy that can be transferred in an inelastic collision is

$$\Delta U_{max} = \frac{1}{2}\left(\frac{m_t m_i}{m_t + m_i}\right)v_i^2 \cos^2\theta \qquad (67)$$

For a collision of equal masses, this reduces to

$$m_t = m_i \rightarrow \Delta U_{max} = \frac{m v_i^2}{4}\cos^2\theta$$

Note that the above energy is equal to one half of the incident particle's kinetic energy in the direction parallel to the centerline. For the collision of unequal masses, the fraction transferred approaches that of the total kinetic energy in the parallel direction:

$$m_t > m_i \rightarrow \Delta U_{max} = \frac{m_i v_i^2}{2}\cos^2\theta \qquad (68)$$

It should be remembered that these processes are reversible, i.e., internal energy can be transferred into kinetic energy during a collision.

In practice, the spectrum of possible internal energy states is not continuous. Complex molecules have a more continuous spectrum, however, since more combinations of rotational, vibrational, and electronic states are allowed, a larger number of closely spaced states would appear as continuous.

There are a large number of possible processes that can occur upon collision. They include electron impact ionization, excitation, relaxation, dissociation, electron attachment, and ion-neutral collisions. Each of the processes can be quantified with a cross-sectional area that is proportional to the rate at which the collisional process takes place. Gas-phase reactions, which are often consequences of these inelastic collisions, are discussed in the following section.

PRINCIPLES OF PLASMA PROCESSING
Course Notes: Prof. J. P. Chang

Part B3: ATOMIC COLLISIONS AND SPECTRA

I. ATOMIC ENERGY LEVELS

Atoms and molecules emit electromagnetic radiation or photons when the electrons or nuclei undergo transitions between various energy levels of the atomic or molecular system. Detailed theory of radiation requires quantum electrodynamics to fully describe the interaction between materials and electromagnetic radiation, so it will not be detailed in this lecture. However, a small set of rules will be discussed to allow the study of the basic physics of radiation.

First we will consider the interaction between electrons in the bound states of atoms and electromagnetic radiation. Atoms emit electromagnetic radiation or photons when their bound electrons undergo transitions between various energy levels of atomic system. Each atomic system has its unique energy levels determined by the electromagnetic interaction among various bound electrons and nucleus.

Calculation of atomic energy levels requires solving a Schrodinger equation for a many particle system (nucleus and electrons) and there is no exact solution available except for the simplest atomic system, i.e., the hydrogen atom. Many approximation methods were developed to calculate the atomic energy levels, currently, the energy levels of many atomic systems are identified and tabulated in the form of Grotrian diagram. One way to designate the various energy levels in the Grotrian diagram is called LS (also known as Russell-Saunders) coupling. However, it should be noted that the LS coupling scheme does not necessarily specify each energy level uniquely, thus one should be careful about using LS coupling scheme. According to the LS coupling scheme, each state is denoted by its orbital angular momentum and the spin state along with each electron's configuration state.

For example, the ground state neutral helium described by LS coupling scheme is $(1s)^2$ $^1S_0$. $(1s)^2$ denotes that two electrons occupy the 1s state, and $^1S_0$ is the spectroscopy term where the superscript 1 denotes the net spin state is singlet, S denotes that the total orbital angular momentum is 0, and the subscript 0 denotes that the total angular momentum is 0 (J=L+S=0).

Another example: the ground state neutral carbon and oxygen described by LS coupling scheme are $(1s)^2(2s)^2(2p)^2$ $^3P_0$ and $(1s)^2(2s)^2(2p)^4$ $^3P_2$. Here we have 2 electrons in the 1s state, 2 electrons in the 2s state, and 2 or 4 electrons in the 2p state. In addition, the net spin state is triplet (superscript 3), the orbital angular momentum state is P (angular momentum quantum number is 1) and the total angular momentum state is 0 or 2 (for less than half filled orbital, J=L-S=0; for half filled orbital or more than half-filled orbital, J=L+S= 1 + 1 =2).

In summary, the designation of atomic energy levels can be done using the spectroscopic designation of an atomic state:

$$X \ I \ n \ ^{2S+1}L_J$$

where X is the element symbol,

I is the ionization state (I: not ionized, II singly ionized, II: doubly ionized, etc.),

n is the principal quantum number,

2S+1 is the multiplicity (S=0: singlet; S=½: doublet; S=1: triplet, etc),

L is the total orbital angular momentum (S, P, D, F, G for L = 0, 1,2,3,4), and

J = L + S is the total electronic angular momentum.

The atomic energy levels of Na is shown in Fig. 1 as an example.

II. ATOMIC COLLISIONS

In a homogeneous plasma, energetic electrons undergo collision with the neutrals to generate excited neutrals, atoms, free radicals, ions, and additional electrons. These electron collision processes make the plasma chemistry complex and interesting. Due to the large mass difference, the electron-particle collision can be viewed as an elastic collision process, as shown in Fig. 2. Several other electron-atom collision processes are listed:

1. Excitation processes

a) Electron impact ionization (Fig. 3):

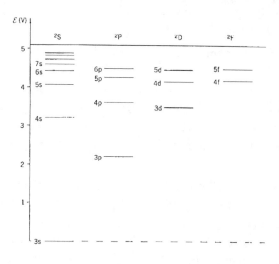

Fig. 1. Atomic energy levels for the central field model of an atom (e.g., Na).

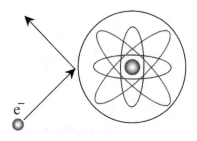

Fig. 2. An electron – atom elastic collision.

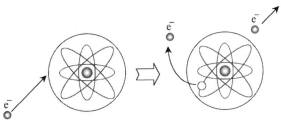

Fig. 3. Electron impact ionization.

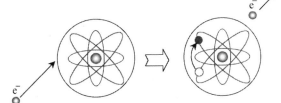

Fig. 4. Electron impact excitation.

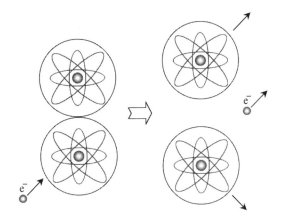

Fig. 5. Electron impact dissociation.

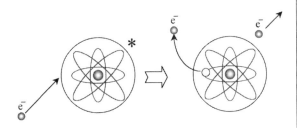

Fig. 6. Electron metastable ionization.

$$e^- + A \rightarrow e^- + e^- + A^+$$

Electrons with sufficient energy can remove an electron from an atom and produce one extra electron and an ion. This extra electron can again be accelerated to gain enough energy and ionize another atom. This multiplication process leads to a continuous generation of ionized species and the plasma is sustained. The ionization processes generally have the highest energy barriers, on the order of 10 eV.

b) Electron impact excitation (Fig. 4):

$$e^- + A \rightarrow e^- + A^*$$

Electrons with sufficient energy can also excite the electrons of an atom from the lower energy level to a higher energy level. This process produces an excited neutral species whose chemical reactivity towards the surface could be quite different from the ground state atoms. The threshold energy needed to produce excited species can vary greatly, depending on the molecule and the type of excitation.

Some excited atoms have very long lifetimes (~ 1-10 msec) because the selection rules forbid its relaxation to the ground state. These excited atoms are thus called *metastables*. All noble gases have metastable states.

c) Electron impact dissociation of diatomic molecules (A_2) (Fig. 5):

$$e^- + A_2 \rightarrow e^- + A + A$$

Electrons with sufficient energy can also break the chemical bonds of a molecule and produce atomic species. These atomic species could gain enough energy and be at a higher energy level than the ground state atoms. Dissociative processes usually have lower threshold energies than ionization processes. Dissociative threshold energies vary from 0 to above 10 eV, depending upon the strength of the bond that is broken and the mechanism by which the process occurs.

This process is mostly responsible for the production of chemically active radicals in most of the plasmas.

d) Electron metastable ionization (Fig. 6):

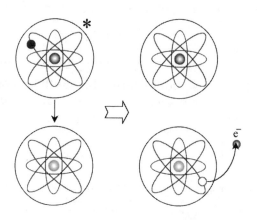

Fig. 7. Penning Ionization.

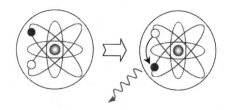

Fig. 8. De-excitation.

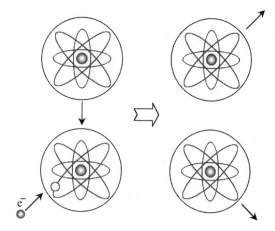

Fig. 9. Three-body recombination.

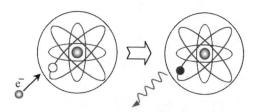

Fig. 10. Radiative recombination (3-body process).

$$e^- + A^* \rightarrow e^- + e^- + A^+$$

Electrons with sufficient energy can also remove an electron from a metastable atom and produce one extra electron and an ion. Since the metastable atom is already excited, less energy is required here to ionize the atom.

e) Metastable-neutral ionization (Fig. 7):

$$A^* + B \rightarrow A + e^- + B^+$$

Metastable atom can collide with a neutral and ionize it if the ionization energy of the neutral (B) is less than the excitation energy of the metastable (A^*). This is also called the *Penning Ionization* process.

2. Relaxation and Recombination Processes

a) De-excitation (Fig. 8):

$$A^* \rightarrow A + h\nu$$

The excited states of atoms are usually unstable and the electron configuration can soon return to its original ground sate, accompanied by the emission of a photon with a specific energy that equals the energy difference between the two quantum levels.

b) Electron-ion recombination (Fig. 9):

$$e^- + A^+ + A \rightarrow A^* + A$$

For electron-ion recombination, a third-body must be involved to conserve the energy and momentum conservation. Abundant neutral species or reactor walls are ideal third-bodies. This recombination process typically results in excited neutrals.

c) Radiative recombination (Fig. 10):

$$e^- + A^+ \rightarrow A + h\nu$$

Photon can also be generated during the coalescence process of recombination. This is also a three-body recombination process, since the two-body coalescence is highly unlikely from the standpoint of energy and momentum conservations.

d) Electron attachment (Fig. 11):

$$e^- + A \rightarrow A^-$$

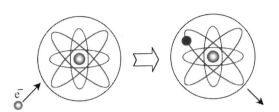

Fig. 11. Electron attachment.

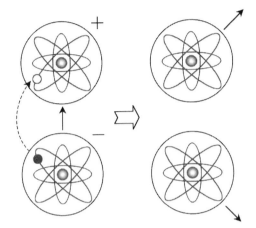

Fig. 12. Ion-ion recombination.

Electron can attach to an electronegative atom to form a negative ion, for example, a halogen atom or an oxygen atom. Complex gas molecules such as SF_6 can also undergo dissociative attachment to form negative SF_5^- ions. This could also be a three-body recombination process.

e) Ion-ion recombination (Fig. 12):

$$A^+ + A^- \rightarrow A + A$$

With negative ions generated, positive ions and negative ions can collide with finite (usually small) probabilities. In ion-ion recombination, one electron transfers and two neutrals are formed.

III. ELASTIC COLLISIONS

1. Coulomb collisions

In general, collisions between ion-ion, ion-electron, and electron-electron are all Coulombic collision. The coulomb potential is:

$$U(r) = \frac{q_1 q_2}{4\pi\varepsilon_0 r} \quad (1)$$

Following the derivation above, the differential collision cross-section can be determined to be:

$$I(v_R, \phi) = \left(\frac{b_0^2}{4\sin^2\frac{\phi}{2}} \right)^2 \quad (2)$$

Note: this is the Rutherford Back Scattering (RBS) cross-section where b_0 is the distance of the closest approach.

$$b_0 = \frac{q_1 q_2}{4\pi\varepsilon_0 W_R} = \frac{Z_1 Z_2 e^2}{4\pi\varepsilon_0 \left(\frac{1}{2} m_R v_R^2 \right)} \quad (3)$$

From this analysis, Coulombic scattering could lead to a single large-angle scattering (less likely) or cause a series of small-angle scatterings.

2. Polarization scattering

With a point charge, q_o, approaches an atom whose atomic radius is a with a point positive charge of q and a uniform negative charge cloud $-q$,

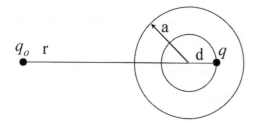

Fig. 13. Polarization of an atom (atomic radius a) by a point charge q_o.

Table 1. Relative polarizability.

Species	**α_R**
H	4.5
C	12
N	7.5
O	5.4
Ar	11
CCl_4	69
CF_4	19
CO	13
CO_2	17
Cl_2	31
H_2O	9.8
NH_3	14.8
O_2	10.6
SF_6	30

the point charge can polarize the atom by displacing the uniform charge cloud through quasistatic interactions. The induced electric field due to a small displacement, d, around the center of the atom is:

$$E_{ind} = -\frac{qd}{4\pi\varepsilon_0 a^3} \qquad (4)$$

The induced dipole is therefore:

$$P_{ind} = qd = \frac{q_o a^3}{r^2} \qquad (5)$$

The attractive potential due to the incoming charge q_o is:

$$U(r) = -\frac{q_o^2 a^3}{8\pi\varepsilon_0 r^4} \qquad (6)$$

The polarizability in this simple atomic model is: $\alpha_p = a^3$, and the relative polarizability is:

$$\alpha_R \equiv \frac{\alpha_p}{a_0^3} \qquad (7)$$

Table 1 summarizes the relative polarizability of several atomic species. Note again that a_o is the Bohr radius.

If the impact parameter, h is small enough, i.e., smaller than the critical impact parameter, h_L, the particle will be captured by the atom during this type of collision. This critical impact parameter is:

$$h_L = \left(\frac{\alpha_p q_o^2}{v_R^2 \pi\varepsilon_0 m_R}\right)^{\frac{1}{4}} \qquad (8)$$

$$m_R = \frac{m_1 m_2}{m_1 + m_2} \qquad (9)$$

$$v_R = |v_1 - v_2| \qquad (10)$$

The Langevin or capture cross-section can thus be determined as:

$$\sigma_L = \pi h_L^2 \qquad (11)$$

IV. INELASTIC COLLISIONS

1. Constraints on electronic transitions

Atoms emit electromagnetic radiation (photons) when the electrons undergo transitions between various energy levels. Since the typically radiation time is on the order of 1 ns, much shorter than the characteristic time between collisions, which are on the order of 100 ns, the excited states will generally be de-excited by electric dipole radiation rather than by collision.

However, not every transition occurs as frequently as others do. The most frequent transition between various energy levels is the electric dipole transition and the following conditions should be satisfied for the electric dipole transition. The general rule of thumb includes:

Energy conservation: the energy of emitted radiation (photons) should be equal to the energy difference between the upper energy level and the lower energy level, $h\nu = E_i - E_j$, where h is the Planck's constant, ν is the frequency of the emitted photon, E_i is the energy of the upper level the electron occupies prior to the transition, and E_j is the energy of the lower level the electron occupies after the transition.

Selection Rules: during the electric dipole transition, the following changes for angular momentum state need to occur:

- Change in the orbital angular momentum state: $\Delta L = 0, \pm 1$

  (0 is not allowed for a transition involving only one electron)

- Change in the spin angular momentum state: $\Delta S = 0$

- Change in the total angular momentum state: $\Delta J = 0, \pm 1$

  (except that J=0 to J=0 transition is strictly forbidden).

From the selection rule, the energy levels of He, can be shown divided into singlet (para-helium) and triplet (ortho-helium) states, since the transitions between them are forbidden. Since L=0 → L=0 is forbidden, the 2^1S and 2^3S states are metastables (Fig. 14). A more detailed Grotrian diagram is included at the end of this section.

It is noted that the selection rules are not perfect, unlike energy conservation. For example, the very intense mercury resonance line at 253.7 nm is due to the transition from $^3P_1 \rightarrow {}^1S_0$.

If the above two conditions are satisfied, the electrons can spontaneously undergo transition from the upper energy level, i, to the lower energy level, j, with a certain probability per unit time. This

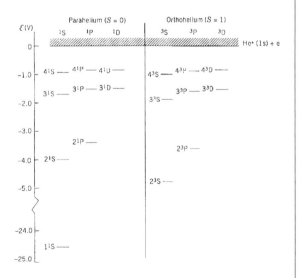

Fig. 14. Atomic energy levels of He, showing the division into singlet and triplet states.

probability is called the transition probability for spontaneous emission (also known as Einstein A coefficient) and can be easily found in the literature for many transitions. For example, the transition probability of hydrogen atom between 2p state and 1s state is 6.28×10^8 sec^{-1}.

2. Identification of atomic spectra

Based on the above discussion, we can now understand the essential features of atomic spectrum and obtain some useful information about the plasma system. As shown in Fig. 15, atomic spectrum usually consists of a number of very sharp lines on the constant background. When the spectrum is measured, the first task is to identify every emission line in the spectrum. This is done by comparing the wavelength of the emission lines with the energy differences between two electronic levels using the published spectral database (NIST database). It is noted that in some cases even this first step is not very straightforward and requires additional consideration. Once this step is completed, we can have at least two (maybe more) very useful information about the plasma. They are:

• Identification of existing atomic species in the plasma.

• Identification of certain excited atomic states and their density in the plasma.

Later, we will learn how to use the information to understand the plasma state.

Light emission is a major characteristic of plasmas. To emit the light, the atoms in the plasma have to be in the excited states. There are two different ways to excite the atoms in the plasma to the excited states. The first one is to use the kinetic energy of the particles in the plasma (in particular electrons) and to transfer this energy to the atoms in the ground state (or another excited state) by collision. This process is called collisional excitation. The second process is to use the energy of the photons and to transfer their energy to the atoms by absorption of photons. This process is called radiative excitation. In most plasma systems, the frequency of the radiative excitation is much smaller than the collisional excitation, thus can be neglected.

Note again that it is not very easy to excite the

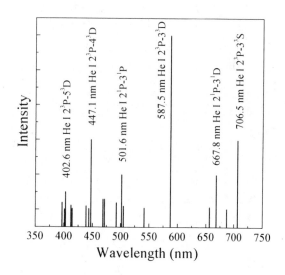

Fig. 15. Emission from a He Plasma.

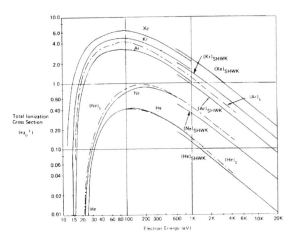

Fig. 16. Ionization cross-section of noble gases.

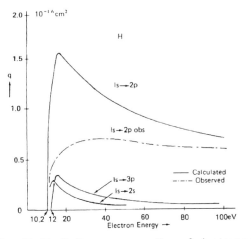

Fig. 17. Excitation cross-section of electrons in hydrogen.

ground state electron in the atom to an excited state. The energy required for this excitation is fairly large. For example, in hydrogen atom, a minimum energy of 10.2 eV is required to move the electron from the ground state (1s) to the lowest excited state (2p) from which atom can emit the photons. An ionization process requires more energy than the excitation process (for example, the ionization potential of hydrogen is 13.6 eV). The ionization cross-sections of several noble gases and the excitation corss-section for H are shown in Fig. 17 and Fig. 18 as examples.

Once the atoms in the plasma are excited above the ground state, it will eventually be de-excited to the ground state. There are three different ways to de-excite the atoms in the plasma. The first one is the spontaneous emission when the electron in the excited level makes a transition to the ground level or another excited level without any external influence. As briefly mentioned earlier, the time scale for this de-excitation is very short if the transition is electric dipole transition, on the order of 10^{-8} sec to 10^{-7} sec. In this case, the energy conservation is satisfied by emitting the photon whose energy is equal to the energy difference between the initial state and the final state. In many plasma systems, this is the most important de-excitation mechanism. On the other hand, the electron in the excited level also makes a transition if there are other photons around the excited atoms. This process is called stimulated emission. Though the stimulated emission is the key element for the laser, in most plasma system, the stimulated emission can be neglected. The third process for the de-excitation is the inverse process of the collisional excitation and is called collisional de-excitation. In collisional de-excitation, the colliding particles will gain energy from the excited atoms into their kinetic energy. The importance of collisional de-excitation is a function of plasma density and electron temperature and it varies for various excited states.

3. A simplified model for emission

To simplify the discussion, we will make a number of assumptions on our plasma system.

1. Plasma density (n_e) is uniform throughout the volume.

2. Electron energy distribution is Maxwellian and

its temperature is given as T_e.

3. Our plasmas are made of hypothetical atoms that have only 4 energy levels, ground state, first and second excited state and ionized state.

4. The rate of collisional de-excitations are small compared to the rate of spontaneous emission, thus will be neglected.

5. The system is in steady state.

As shown in Fig. 18, E_o, E_1, E_2, and E_i are the ground state, excited state 1, excited state 2, and the ionized state. R_1, and R_2 are rate of collisional excitation from the ground state, and A_{1o}, A_{2o}, and A_{21} are rate of spontaneous emission (Note that they are also called Einstein A coefficient). R_1, and R_2 can be calculated using the collisional cross-sections:

$$R_1 = n_e n_0 < \sigma_1 v >$$

$$R_2 = n_e n_0 < \sigma_2 v >$$

Remember $<\sigma v>$ is the collision rate averaged over the MBD.

From these rates, the equations governing the density of each state can be determined:

$$\frac{dn_0}{dt} = -n_e n_0 \left(< \sigma_1 v > + < \sigma_2 v > \right) + n_1 A_{10} + n_2 A_{20}$$

$$\frac{dn_1}{dt} = n_e n_0 < \sigma_1 v > + n_2 A_{21} - n_1 A_{10}$$

$$\frac{dn_2}{dt} = n_e n_0 < \sigma_2 v > - n_2 \left(A_{21} + A_{20} \right)$$

In steady state, the time derivatives in the LHS are zero, and we have two independent equations with five unknowns (n_e, n_0, n_1, n_2, T_e).

From charge quasi-neutrality and particle conservation, we have one more equation:

$$n_e + n_0 + n_1 + n_2 = n_g$$

where n_g is the gas density without the plasma.

During spontaneous emission, the excited state 1 and 2 will emit photons at the following frequencies:

$$h v_{10} = E_1 - E_0$$

$$h v_{20} = E_2 - E_0$$

$$h v_{21} = E_2 - E_1$$

If we can measure the number of photons emitting

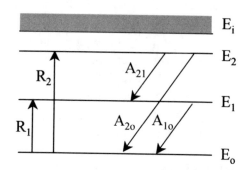

Fig. 18. Energy diagram of an atom with limited energy levels.

at these frequencies, we can then determine the density of excited states (n_1 and n_2) independently:

$$\frac{\text{\# photons at } \nu_{10}}{\text{time}} = n_1 A_{10} \cdot \text{volume}$$

$$\frac{\text{\# photons at } \nu_{20}}{\text{time}} = n_2 A_{20} \cdot \text{volume}$$

Now we have three unknowns (n_e, n_0, T_e) for three equations:

$$n_e n_0 < \sigma_1 v > + n_2 A_{21} - n_1 A_{10} = 0$$

$$n_e n_0 < \sigma_2 v > - n_2 \left(A_{21} + A_{20} \right) = 0$$

$$n_e + n_0 + n_1 + n_2 = n_g$$

Therefore n_e, n_0, and T_e can be calculated.

Unfortunately the situation in real systems is very different from this simplified model. Thus, the use of plasma spectroscopy alone may not provide the enough information about the plasma system that we want to know. However, we can still obtain some very valuable information on our plasma system from plasma spectroscopy.

Fig. 19. He Grotrian Diagram

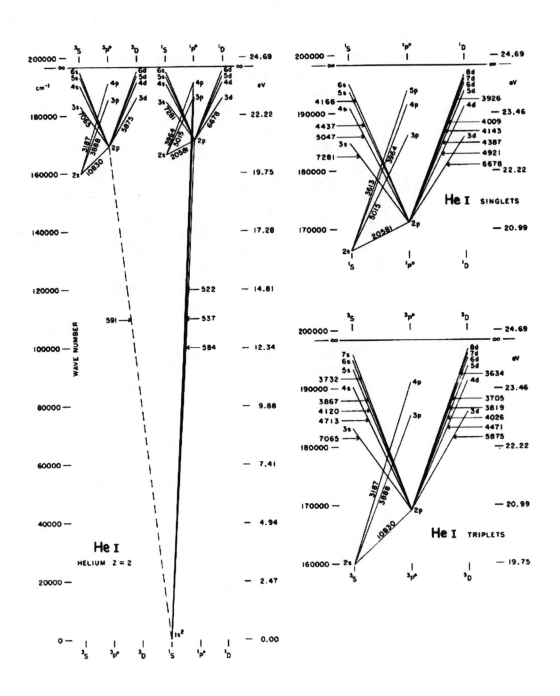

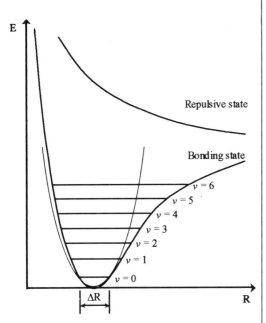

Fig. 1. Electronic states of a molecule.

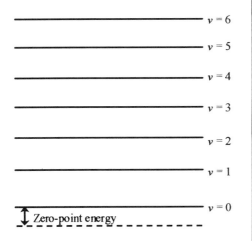

Fig. 2. Vibrational energy levels.

PRINCIPLES OF PLASMA PROCESSING
Course Notes: Prof. J. P. Chang

PART B4: MOLECULAR COLLISIONS AND SPECTRA

Like atoms, molecules emit photons when they undergo transitions between various energy levels. However, in molecules, additional modes of motion are possible. They are rotation and vibration of molecules. Note that little energy is coupled through vibrational and rotational states as these processes are inefficient. This poor coupling can be thought of as a momentum limitation; i.e., the low mass, high velocity electron cannot excite these states in which momentum must be transferred to an atom. In a typical electron excitation of rotational states, a single quantum is transferred. As a quantum for rotational states are of the order of 1 milli-eV, little energy is transferred. In electron impact excitation of vibrational states, again typically only a single quantum is transferred. The vibrational quanta energy is of the order of 0.1 eV. An exception is the vibrational excitation of molecules in which the electron attaches to form a negative ion which is short lived. The negative ion has a different interatomic spacing, therefore, when the electron subsequently leaves, the molecule finds itself with a bond length that differs from the neutral state. The bond acts as a spring converting the energy into many quanta of vibrational energy. A typical chemical bond is of the order of 4-5 eV; therefore in the discharges used in microelectronics processing, the excitation of rotational and vibrational states are typically not significant. Nevertheless, the energy levels of molecules are further complicated due to these additional modes of motion.

I. MOLECULAR ENERGY LEVELS

The molecular energy level can be represented by:

$eE = eE_e + eE_v + eE_J$, where

eE_e is the electronic energy level

eE_v is the vibrational energy level = $hv\left(v+\frac{1}{2}\right)$

$eE_J = eE_r$ is the rotational energy level = $J(J+1)\dfrac{h^2}{8\pi^2 I}$

Note that v is the vibrational quantum number, and J is the rotational quantum number.

1. Electronic energy level

For diatomic molecules, the electronic states are specified by the total orbital angular momentum, Λ, along

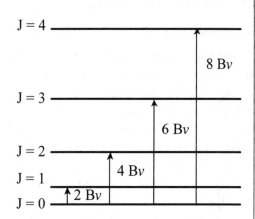

Fig. 3. Rotational energy levels.

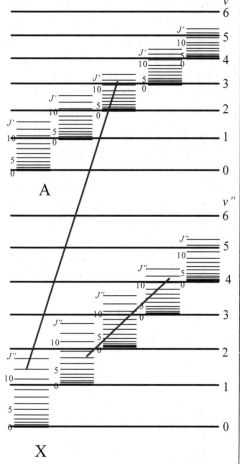

Fig. 4. Vibrational and rotational energy levels.

the internuclear axis; and for Λ = 0, 1, 2, 3, symbols $\Sigma, \Pi, \Delta, \Phi$ are used. Note that all but the Σ states are doubly degenerated.

The total electron spin angular momentum S is used to determined the multiplicity, 2S+1, and is written as a prefixed superscript, as for the atomic states.

Analogous to the atomic LS coupling for atoms, another quantum number denoted as $\Omega=\Lambda+\Sigma$ is used as a subscript. Note that the allowed values of Σ are S, S-1, S-2, …., -S.

To complete the description for molecular spectroscopic terms, note that g (*gerade* or even) or u (*ungerade* or odd) subscripts denote whether the wavefunction is symmetric or antisymmetric upon inversion with respect to its nucleus. Superscripts of + and − are used to denote whether the wave function is symmetric or antisymmetric with respect to a refection plane through the internuclear axis.

So, the spectroscopic designation of a molecular state is:

$$^{2S+1}\Lambda^{+(-)}_{\Omega g(u)}$$

For example, the ground state of H_2 and N_2 are singlets, $^1\Sigma_g^+$, while the ground state of O_2 is a triplet, $^3\Sigma_g^-$.

A typical set of states in a diatomic molecule is in Fig. 1: the lower curve is the ground electronic state in which the lowest energy is indicated by the x axis. In the lowest state, the molecule vibrates with the interatomic distance varying over ΔR. Also indicated are the next few higher vibrational states and their interatomic ranges. The rotational states can also be added to the vibrational states. The upper curve is an excited state that is repulsive. Note that an excited state can also be bonding with a minimum energy.

2. Vibrational energy level

For a harmonic oscillator, the vibrational frequency $\propto \sqrt{\dfrac{k}{m}}$. For a diatomic molecule, the vibrational frequency $\propto \sqrt{\dfrac{k}{m_R}}$, where m_R is the reduced mass of the system.

The vibrational energy level is $eE_v = h\nu\left(v+\dfrac{1}{2}\right)$.

Therefore, the energy spacing is almost the same, but the spacing does decrease with increasing vibrational quantum number due to the anharmonic motion of the

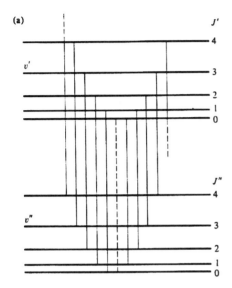

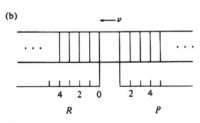

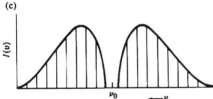

Fig. 5. HCl: (a) Details of allowed vibrational and rotational transitions, (b) spectrum lines, (c) intensity distribution.

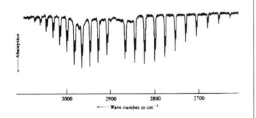

Fig. 6. Actual infrared absorption spectrum of HCl. The fine splitting is due to $H^{35}Cl$ and $H^{37}Cl$ isotopic shift.

molecule (Fig. 2). Note that the lower energy level is typically labeled as $v^{''}$ and higher energy level is labeled as $v^{'}$, as shown later in Fig. 4.

3. Rotational energy level

For a simple dumbbell model for diatomic molecules, the moment of inertia is $I = m_R r^2$. The rotational energy level is:

$$E_J = J(J+1)\frac{h^2}{8\pi^2 I} \equiv B_v J(J+1)$$

Therefore, the energy spacing increases with increasing rational quantum number (Fig. 3). Again, the lower energy level is typically labeled as $J^{''}$ and higher energy level is labeled as $J^{'}$, as shown in Fig. 4, where the details of the allowed vibrational and rotational transitions, spectrum lines, and intensity distribution. Note that X denotes the ground state, while A represents an excited state.

Figure 5 shows the theoretical infrared absorption spectrum of a diatomic molecule, HCl: (a) the allowed vibrational and rotational transitions, (b) the measured spectrum lines, and (c) the intensity distribution. Note that the P branch represents the transitions corresponding to $\Delta J = -1$, while the R branch represents the transitions corresponding to $\Delta J = +1$. The Q branch is missing since the transition of $\Delta J = 0$ is forbidden. The actual experimental result is shown in Fig. 6, while the fine splitting is due to the isotopic shift of $H^{35}Cl$ and $H^{37}Cl$.

II. SELECTION RULE FOR OPTICAL EMISSION OF MOLECULES

For practical applications, the following (approximate) selection rules are given for molecular transitions:

Change in orbital angular momentum: $\Delta\Lambda = \pm 1$
Change in spin angular momentum: $\Delta S = 0$

The selection rule for $v^{'}$ to $v^{''}$ is: $\Delta v = \pm 1$
The selection rule for $J^{'}$ to $J^{''}$ is: $\Delta J = \pm 1$

In addition, for transitions between Σ states, the only allowed transitions are $\Sigma^+ \rightarrow \Sigma^+$ and $\Sigma^- \rightarrow \Sigma^-$; and for homonuclear molecules, the only allowed transitions are $g \rightarrow u$ and $u \rightarrow g$.

III.   ELECTRON COLLISIONS WITH MOLECULES

The interaction time of an e⁻ with a molecule is:

$$t_c \sim 10^{-16} - 10^{-15} \text{ s}$$

The typical time for a molecule to vibrate is:

$$t_{vib} \sim 10^{-14} - 10^{-13} \text{ s}$$

The typical time for a molecule to dissociate is:

$$t_{diss} \sim t_{vib} \sim 10^{-14} - 10^{-13} \text{ s}$$

The typical transition time for electric dipole radiation is:

$$\tau_{rad} \sim 10^{-9} - 10^{-8} \text{ s}$$

The typical time between collision in a low pressure plasma is

$$\tau_c$$

These time scales are:

$$t_c \ll t_{vib} \sim t_{diss} \ll \tau_{rad} \ll \tau_c$$

1.   Frank-Condon principle

Since $t_c \ll t_{vib,}$ electronic excitations are indicated by vertical transitions in Fig. 7 as the interatomic distance cannot change in the time scale of the excitation. Such a process is sometimes called a Frank-Condon or adiabatic transition.

Since $\tau_{rad} \gg t_{diss,}$, if the energetics permit, the molecule will dissociate instead of de-exciting to the ground state.

It should be noted that only certain energies can be adsorbed which correspond to the spacings indicated; however, the distribution in interatomic spacing as the molecules vibrate result in a broadening of the acceptable excitation energies. Photoelectron excitations occur in a manner similar to this; however they have an additional constraint of spin conservation. In electron impact, spin conservation is not important as the process can be treated as a three body event. Note that excited states can be short-lived or may be metastable. Various electronic levels have the same energy in the unbound limit ($R \rightarrow \infty$).

2.   Dissociation

Shown in Fig. 8 are various processes which lead to dissociation in molecules.

$$e^- + AB \rightarrow A + B + e^-$$

Processes b-b' result in the excitation to a state in which

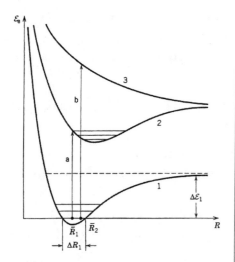

Fig. 7. Frank-Condon or adiabatic transition.

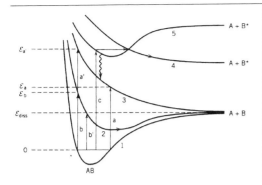

Fig. 8. Dissociation processes.

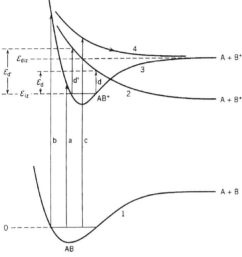

Fig. 9. Dissociative ionization and dissociative recombination processes.

the excited molecule is not stable. This results in the production of A+B with the excess energy being converted into translational energy of the molecular fragments. Excitations to curve 2 with lower energies result in a bonded electronically excited state. Excitation to curve 3 which is repulsive is indicated by processes a-a'. These excitation result in the production of A+B. Excitation c indicates the excitation to an excited state which is stable. This state can relax by the emission of a photon to curve 3 resulting in dissociation or by curve crossing to a repulsive state (curve 4) again resulting in dissociation. Note in the latter process, B* is produced in an electronically excited state.

3. Dissociative ionization

Figure 9 are processes associated with ionization and dissociative ionization.

$$e^- + AB \rightarrow AB^+ + 2e^-$$

Note that curve 2 represents a stable molecular ion state (AB^+). This state can undergo dissociative recombination to produce fast and excited neutral fragments.

$$e^- + AB \rightarrow A + B^+ + 2e^-$$

A repulsive ion state, curve 4 is also shown which always results in fragmentation after ionization.

4. Dissociative recombination

The electron collision illustrated in Fig. 9 as d and d' represents the capture of the electron leading to the dissociation of the molecule. Thus, this process is called dissociative recombination.

$$e^- + AB^+ \rightarrow A + B^*$$

5. Dissociative electron attachment

Depending upon the dissociation energy and the electron affinity of B, the dissociative electron attachment can be categorized into autodetachment, dissociative detachment, electron dissociative attachment, and polar dissociation:

$$e^- + AB \rightarrow AB^-$$
$$e^- + AB \rightarrow AB^- \rightarrow A + B^-$$
$$e^- + AB \rightarrow A^+ + B^- + e^-$$

Figure 10 shows a number of examples of electron attachment processes for molecules: (a) the excitation to a repulsive state requires electron energies greater than the threshold energy, (b) the attachment requires little electron energy and can result in a stable negative ion or fragmentation, (c) the capture of a slow electron to a

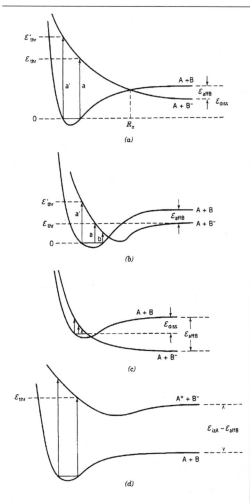

Fig. 10. Electron attachment processes illustrating the capture of electron into: (a) a repulsive state, (b) an attractive state, (c) repulsive state (with slow electrons), and (4) polar dissociation.

repulsive state results in the formation of a negative ion that is a fragment. (d) the excitation to a neutral excited state with sufficient energy that a positive and negative ion are simultaneously formed, i.e., polar dissociation.

6. Electron impact detachment

Electron-negative ion collision can result in the detachment of the electron to yield a neutral and one additional electron. Here the electron affinity of the negative ion plays an important role.

$$e^- + AB^- \rightarrow AB + e^- + e^-$$

7. Vibrational and rotational excitation

Electrons with sufficient energy can excite molecules into higher vibrational and rotational energy levels. This is typically a two-step process, where electron is first captured (a negative ion forms) and then detached to generate a vibrationally excited molecule.

$$e^- + AB \, (v{=}0) \rightarrow AB^-$$
$$AB^- \rightarrow AB \, (v{>}0) + e^-$$

IV. HEAVY PARTICLE COLLISIONS

The collisions between ion-ion, ion-neutral, and neutral-neutral are heavy particle collisions. These species all have much lower temperatures compared to the electrons, thus move much slower compared to the electrons. The important heavy particle collisions are:

1. Resonant and non-resonant charge transfer

Resonant charge transfer is important in producing fast neutrals and slow ions, that would modify the overall chemical reactivity of plasma towards the surface.

$$A^+ + A \rightarrow A + A^+$$

Non-resonant charge transfer can take place between unlike atoms/molecules or between an atom and a molecule.

$$A^+ + B \rightarrow A + B^+$$

Figure 11 shows the non-resonant charge transfer between N^+ and O, while several non-resonant charge transfer reactions between oxygen molecule and atom are important in an oxygen plasma.

2. Positive and negative ion recombination

As discussed in Atomic Collisions and Spectra, ion-ion recombination is a type of charge transfer and can be the dominant mechanism for the loss of negative ions in a low pressure electronegative plasma.

$$A^- + B^+ \rightarrow A + B^*$$

3. Associative detachment

The associative detachment process is shown in Fig. 12. Depending upon the energy level of AB⁻, the dissociation process varies.

$$A^- + B \rightarrow AB + e^-$$

4. Transfer of excitation

As discussed in Atomic Collisions and Spectra, transfer of excitation can take place in the plasma, including the Penning ionization.

$$A + B \rightarrow A^+ + B + e^-$$
$$A + B \rightarrow A^* + B$$
$$A + B^* \rightarrow A^+ + B^* + e^- \text{ (Penning ionization)}$$
$$A + B^* \rightarrow AB^+ + e^-$$
$$A + B^* \rightarrow A^* + B$$

5. Rearrangement of chemical bonds

Chemical bond rearrangement can also take place in the plasma, making the composition more complex.

$$AB^+ + CD \rightarrow AC^+ + BD$$
$$AB^+ + CD \rightarrow ABC^+ + D$$

$$AB + CD \rightarrow AC + BD$$
$$AB + CD \rightarrow ABC + D$$

6. Three-body processes

As discussed in Atomic Collisions and Spectra, three body collisions are important processes that conserve the energy and momentum, and allow complex chemical reactions to take place in the plasma gas phase.

a) Electron-ion recombination
$$e^- + A^+ (+e^-) \rightarrow A + (+e^-)$$

b) Electron attachment
$$e^- + A (+M) \rightarrow A^- + (+M)$$

c) Association
$$A^+ + B (+M) \rightarrow AB^+ + (+M)$$

d) Positive-negative ion recombination
$$A^- + B^+ (+M) \rightarrow AB + (+M)$$

V. GAS PHASE KINETICS

The unique chemical reactions that take place in a plasma are almost entirely caused by inelastic collisions between energetic electrons and neutrals of thermal

$$N^+ + O \rightarrow N + O^+$$

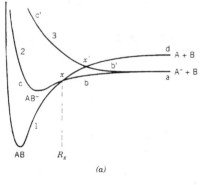

Fig. 11. Nonresonant charge transfer processes.

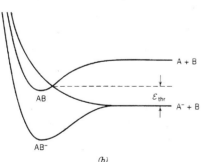

Fig. 12. Associative detachment process: (a) AB⁻ ground state above AB ground state, (b) AB⁻ ground state below AB ground state.

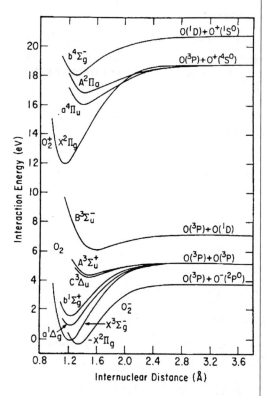

Fig. 13. Simplified energy diagram for O_2.

Table 1. Threshold energy for oxygen excitation.

Reactions	E_{th} (eV)
1) $O_2 + e^- \rightarrow O_2(v) + e^-$, $v = 1..10$	1.95
2) $O_2 + e^- \rightarrow O_2(^1\Delta_g) + e^-$	0.98
3) $O_2 + e^- \rightarrow O_2(b^1\Sigma_g^+) + e^-$	1.64
4) $O_2 + e^- \rightarrow O_2(A^3\Sigma_u^+) + e^-$	4.50
5) $O_2 + e^- \rightarrow O_2(*) + e^-$	6.00
6) $O_2 + e^- \rightarrow O_2(**) + e^-$	8.00
7) $O_2 + e^- \rightarrow O_2(***) + e^-$	9.70
8) $O_2 + e^- \rightarrow O_2^+ + 2\,e^-$	12.20
9) $O_2 + e^- \rightarrow O + O + e^-$	6.00
10) $O_2 + e^- \rightarrow O^- + O$	3.60

energy. The inelastic scattering produces a host of excited states, which then relax and/or interact by collision between particles or by collisions with the walls of the reactor.

An energy level diagram for oxygen, a diatomic molecule, is shown in Fig. 13 to illustrate the complexity of possible gas phase reactions in an oxygen plasma. The electron states of O_2^-, O_2, and O_2^+ are shown. Only attractive states are shown in this simplified energy diagram, though repulsive state do exist. Several attractive states shown here are metastables, including $^1\Delta_g$, $^1\Sigma_g^+$, and $^3\Delta_u$ states of O_2. The threshold energy for oxygen excitation processes is shown in Table 1.

A short list of the reactions that take place in an oxygen plasma is in Table 2 for an analysis: Reactions 1 and 2 involve the inelastic scattering of an electron with neutrals and are characterized by collision cross-sections, while reactions 3-5 are heavy particle collisions and are quantified in terms of rate coefficients.

Table 2. Reactions in an oxygen glow discharge

Reaction	ki	σ, cm^2
1. Ionization: $e^- + O_2 \rightarrow O_2^+ + 2e^-$		2.7×10^{-16}
2. Dissociative attachment: $e^- + O_2 \rightarrow O + O^-$		1.4×10^{-18}
3. Charge transfer: $O^+ + O_2 \rightarrow O_2^+ + O$	2×10^{-11} cm^3/s	
4. Detachment: $O^- + O \rightarrow O_2 + e^-$	3×10^{-10} cm^3/s	
5. Atom recombination: $2O + O_2 \rightarrow 2O_2$	2.3×10^{-33} cm^6/s^2	

The cross-sections can be related to to an effective reaction rate coefficient by:

$$k_i = \int_0^\infty \sqrt{\frac{2E}{m}} \sigma_i(E) f(E) dE \qquad (1)$$

This equation represents an integration over all electron energies of the product of the electron velocity, the collision cross-section, and the electron-energy distribution function. The collision cross-section can be considered the probability that during a collision a certain reaction takes place. It has the units of area to be dimensionally correct; however, this area has only a vague interpretation in terms of the distance at which the particles must approach to react in a specific manner. The collision cross-section is a function of the energy of the electron in most cases. The energy dependence for a number of processes is shown in Fig. 14.

The modeling of an oxygen discharge using MBD is reasonably successful in predicting rate constants for inelastic collisions with a threshold energy below that of the average electron energy. However, for predicting

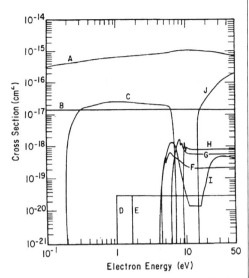

Fig. 14. Elastic and inelastic collision cross sections as a function of energy for electron impact reactions of O_2.
(A) elastic scattering,
(B) rotational excitation
(C) vibration
(D) (E) (F) electronic excitations
(I) dissociative attachment
(J) ionization

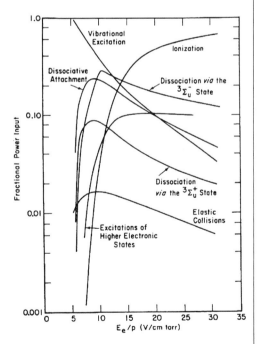

Fig. 15. Fractional power input to elastic and inelastic collisions as a function of E_e/p for oxygen.

higher threshold events, such as ionization, such modeling gives poor results. The failure to model the high energy processes reflects the greater deviation from MBD in the high energy tail region, as the Druyvesteyn model suggests. For the high energy processes, the variation of collision cross-sections with energy and the effects of the electron-velocity distribution must be taken into account.

In addition, the power dissipation in a plasma can be related to the various collisions and their energies,

$$\langle Power \rangle = \frac{2m}{M}\langle E \rangle k_m nN + \sum_j E_j k_j nN + \langle E \rangle k_i nN + \langle E \rangle k_d n \quad (2)$$

where $\langle E \rangle$ indicates the average excitation energy for each process, E_j is the energy loss for the jth process, $k_m = \nu_m/N$ is the rate constant for momentum transfer, k_j is the rate constant for the jth inelastic process, k_i is the ionization rate constant, and k_d is the effective diffusion rate constant. Since the cross-sections for the processes vary with energy, plotting the fractional energy dissipated for rotational, vibrational, dissociation, and ionization processes reveals significant variation in the partitioning of power to different processes as a function of E_e/p. Figure 15 shows this partitioning of energy for an oxygen discharge. It should be noted that less than 1.5% of the power is lost in elastic collisions.

Since oxygen plasma is widely used in the microelectronics industry to ash photoresist, we will use the production of oxygen atoms in a plasma as an example here. The major mechanisms contributing to oxygen atom generation and loss are listed below:

1a. $e^- + O_2 \rightarrow O_2^*(A^3\sum_u^+) + e^- \rightarrow 2O(^3P) + e^-$

1b. $e^- + O_2 \rightarrow O_2^*(B^3\sum_u^-) + e^- \rightarrow O(^3P) + O(^1D) + e^-$

2. $2O + O_2 \rightarrow 2O_2$
3. $3O \rightarrow O + O_2$
4. $O + 2O_2 \rightarrow O_3 + O_2$

Assuming that the electron energy distribution is MBD, and the rate coefficients can be calculated for each of these reactions ($k_1 - k_4$). In addition, a surface recombination coefficient γ is used to account for atomic oxygen loss through interaction with the walls of the reactor. Assuming that the reactor design can be modeled as a plug flow in the tube, the differential mass balance for the reactor can be written as:

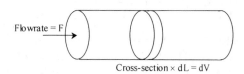

Fig. 16. Plug flow reactor.

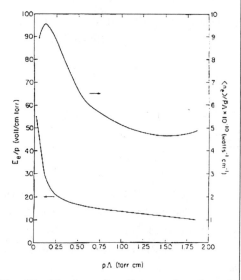

Fig. 17. Electron density as a function of pressure.

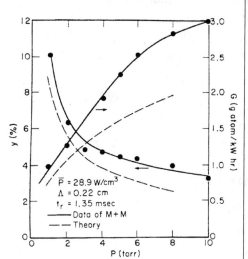

Fig. 18. Conversion and yield vs. pressure.

$$\frac{4Fn_o}{(2n_o-n)^2}\frac{dn}{dV} = -\frac{1}{2R}nv\gamma + 2k_1 <n_e>(n_o-n)$$
$$-2k_2n^2(n_o-n)-2k_3n^3-2k_4n(n_o-n)^2 \tag{3}$$

where n_o is the total number density of gas particles, n is the number density of atomic oxygen, F is the flow rate, V is the reactor volume, and v is the oxygen atom velocity. The left-hand-side term is the total rate at which atomic oxygen atoms within the differential volume dV are accumulated. The concentration can be determined by integrating V from the inlet of the tube, where the density and composition are known, to the exit of the tube. The first term on the right-hand-side is the rate of O loss by recombination on walls of the reactor to form O_2. The following terms are the rate that atoms are created by electron impact reactions 1a and 1b, and the losses by reactions 2, 3, and 4, respectively. Note that $<n_e>$ is a function of p and Λ (the characteristic diffusion length of the system), as shown in Fig. 17. The above equation can be solved and compared with experimental results as shown in Fig. 18. Note that the total reaction yield (total amount of O produced), G, and conversion (fraction concerted to O), y, are defined as:

$$y = \frac{n}{2n_o-n} \tag{4}$$

$$G = \frac{7\cdot10^6 \, yF}{\langle Power\rangle V} \tag{5}$$

The decrease in conversion with pressure is a result of the reduction of the dissociation rate (reaction 1), since $<n_e>$ and k_1 decrease with increasing pressure. The increased pressure also causes an increase in the homogeneous recombination rate, reactions 2-4, but these are minor losses at these low pressures.

This model only predicts the plasma gas-phase concentration of atomic oxygen, and the calculation of ashing rates is more difficult in that it requires additions to the model for both more complex surface reactions and consideration of additional species in the plasma-phase. For each additional species in the plasma, an additional equation similar to that above must be considered and solved simultaneously.

Information about a Cl_2 plasma is shown in Table 3 and Fig. 19 and 20 as a reference and for the homework problems.

Table 3. Gas-phase reaction mechanisms in a chlorine plasma.

1. The reaction threshold energy and constants of the rate constants are listed for comparison.

$$k = \int_0^\infty f(E) \left(\frac{2E}{m}\right)^{1/2} \sigma(E) dE = A T_e^B \exp\left(\frac{-C}{T_e}\right) \; ; \text{units: k [cm}^3\text{s}^{-1}]; \; T_e \text{ [eV]}$$

where A, B, C and the threshold energy are summarized below:

	Reaction	A	B	C	$E_{threshold}$
(1)	$e^- + Cl_2 \rightarrow e^- + Cl_2$	2.18×10^{-2}	-1.433	16 304.0	0.07
(2)	$e^- + Cl_2 \rightarrow Cl^- + Cl$	2.33×10^{-11}	0.237	9163.8	
(3)	$e^- + Cl_2 \rightarrow Cl + Cl + e^-$	2.11×10^{-9}	0.232	54866.0	2.50
(4)	$e^- + Cl_2 \rightarrow e^- + Cl_2$	9.47×10^{-11}	0.445	113840.0	9.25
(5)	$e^- + Cl_2 \rightarrow Cl_2^+ + 2e^-$	1.02×10^{-10}	0.641	150810.0	11.48
(6)	$e^- + Cl^- \rightarrow Cl + 2e^-$	1.74×10^{-10}	0.575	48883.0	3.61
(7)	$e^- + Cl \rightarrow e^- + Cl^*$	2.35×10^{-5}	-0.953	124040.0	9.00
(8)	$e^- + Cl \rightarrow e^- + Cl$	1.53×10^{-9}	0.183	113280.0	9.55
(9)	$e^- + Cl \rightarrow e^- + Cl$	2.14×10^{-10}	0.189	126890.0	10.85
(10)	$e^- + Cl \rightarrow e^- + Cl$	6.35×10^{-11}	0.187	148090.0	12.55
(11)	$e^- + Cl \rightarrow e^- + Cl$	1.07×10^{-8}	0.075	134280.0	11.65
(12)	$e^- + Cl \rightarrow e^- + Cl$	5.47×10^{-9}	0.073	141 400.0	12.45
(13)	$e^- + Cl \rightarrow e^- + Cl$	3.70×10^{-9}	0.053	146 370.0	12.75
(14)	$e^- + Cl \rightarrow e^- + Cl$	2.00×10^{-7}	-0.235	126 730.0	10.85
(15)	$e^- + Cl \rightarrow e^- + Cl$	5.61×10^{-8}	-0.241	143 350.0	12.15
(16)	$e^- + Cl \rightarrow Cl^+ + 2e^-$	5.09×10^{-10}	0.457	155 900.0	13.01
(17)	$e^- + Cl^* \rightarrow Cl^+ + 2e^-$	9.29×10^{-9}	0.265	47 436.0	3.55
(18)	$Cl_2^+ + Cl^- \rightarrow 2Cl + Cl$	1.00×10^{-7}	0.000	0.0	
(19)	$Cl^+ + Cl^- \rightarrow Cl + Cl$	1.00×10^{-7}	0.000	0.0	
(20)	$Cl^+ + Cl_2 \rightarrow Cl_2^+ + Cl$	5.40×10^{-10}	0.000	0.0	
(21)	$Cl + Cl + M \rightarrow Cl_2 + M$	3.47×10^{-33}	0.000	-810.0	
(22)	$Cl^* + Cl_2 \rightarrow 3Cl$	5.00×10^{-10}	0.000	0.0	

M: Third body.

2. Important spectroscopic information for Cl_2:

(a) Electronic state: $^1\Sigma_g^+$
(b) Vibrational constant: 559.7 cm^{-1}
(c) Vibrational anharmonicity: 2.68 cm^{-1}
(d) Rotational constant: 0.2440 cm^{-1}
(e) Rotation-vibration interaction constant: 0.0015 cm^{-1}
(f) Centrifugal distortion constant: 0.186×10^{-6} cm^{-1}
(g) Interatomic distance: 1.988 Å

Figure 19: Elastic and inelastic collision cross sections as a function of energy for electron impact reactions of Cl₂.

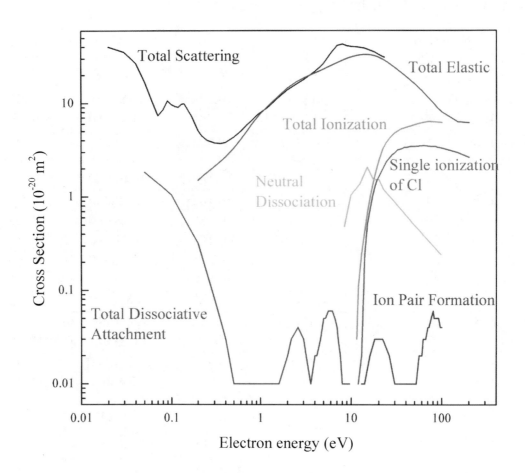

Fig. 20: Potential Energy Diagram of Cl₂.

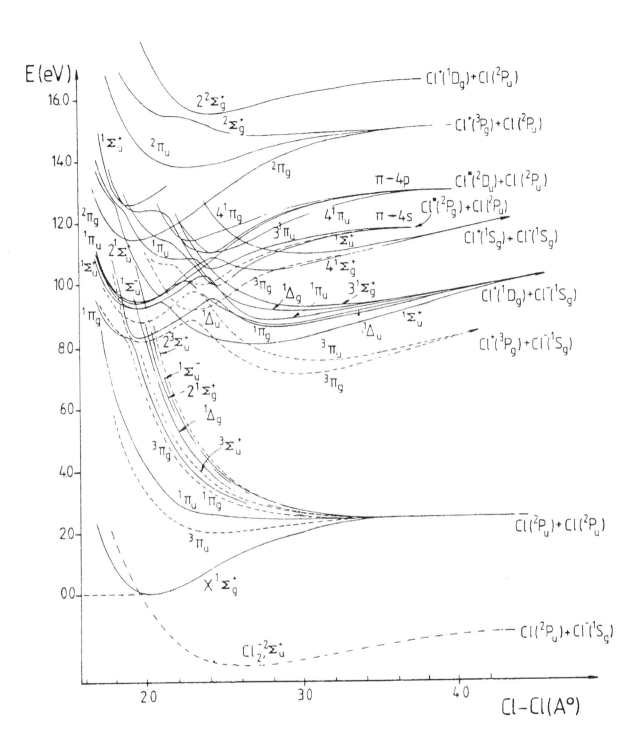

PRINCIPLES OF PLASMA PROCESSING
Course Notes: Prof. J. P. Chang

PART B5: PLASMA DIAGNOSTICS

In order to understand the plasma chemistry, various plasma process diagnostic tools have been developed to quantify the concentrations of various reactive species in the plasma. The basic principles of their operation are described here with examples and their advantages/disadvantages. Keep in mind that a diagnostic technique is no more intelligent than the person who is using it, and it requires careful analysis of the measurement to obtain meaningful data.

I. OPTICAL EMISSION SPECTROSCOPY

1. Optical Emission

Optical emission from a plasma occurs primarily through the electron impact excitation of atoms or molecules to an excited state, followed by a relaxation to a lower energy state releasing a photon containing an energy equal to the difference between these two energy states. Analysis of the photon energy (wavelength of light) and spectral emission information of species can therefore be used to infer the composition of the species that produced it.

Optical emission of atoms is reasonably straight-forward, since only electronic state transitions can occur. Therefore atomic spectra have sharp, nearly mono-energetic, and well-defined peaks corresponding to transitions between various electronic states.

Molecules, however, have a larger number of electronic states, and also have both vibrational and rotational states superimposed upon their electronic states. The small energy differences between the vibrational and rotational states, the broadening of emission energies caused by collisions, and movement of the emitting molecules cause the emission to overlap and form bands rather than sharp emission peaks at easily identified frequencies. Typically, a sharp bandhead is observed and can be used to identify the band, while red or blue shadings, as shown in Fig. 1, are due to the fact that the upper state is less or more tightly bounded than the lower states. Keep in mind that a bandhead or a shading is not always observed.

At higher pressures, collisions broaden the emission energy, but such broadening is not usually observed in the low pressure discharges used in plasma processing. Other

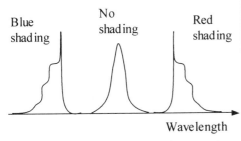

Fig. 1. Signatures of molecular spectra.

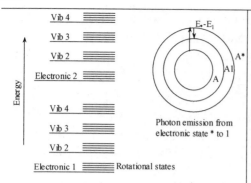

Fig. 2. Schematic of photon emission process

Table 1. Energy separation

Energy level	Energy (eV)	Energy (cm^{-1})
Electronic	0.8-18	6500-145000
Vibrational	0.02-0.6	200-5000
Rotational	0.00001-0.0006	0.1-5

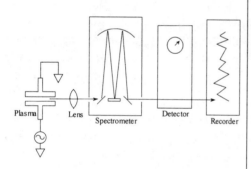

Fig. 3. Schematic drawing of an optical monochromator system for the analysis of photoemission from a plasma.

mechanisms for optical emission are chemiluminescence in which the excess energy from an exothermic chemical reaction emits light. For example, the blue color at the base of a flame is produced chemiluminescence. Emission can also be produced by ion-neutral collisions, sputtering products that are released in an excited state, excited products of electron impact dissociation, absorption of photons and subsequent fluorescence or phosphorescence, collisional relaxation of metastable, etc.

Shown in Fig. 2, the electron impact excitation of the ground state of atom A to an excited state A*, followed by subsequent emission at frequency ν to a lower energy level A1. The usual wavelengths are in the optical band, i.e., $\lambda \sim 2000$–8000Å. The typical energy separation of electronic, vibrational, and rotational transitions is summarized in Table 1.

2. Spectroscopy

Optical emission spectroscopy, which measures the light emitted from a plasma as a function of wavelength, time, and location, is the most commonly used plasma diagnostic probe for microelectronics fabrication processes. The rate at which such transitions occur is determined by the quantum mechanical similarity of the states. This similarity leads to selection rules that indicate the allowed transitions. The energy of the photons emitted by the plasma, therefore, is characteristic of the composition and energy state of species within the plasma. The spectra can be used to analyze both the chemical species that make up the plasma and their state of excitation. Because it is non-intrusive, inexpensive, and can be easily incorporated into an existing plasma reactor, it quickly gains popularity in the microelectronics industry for monitoring the plasma processing. However, the large information content makes the interpretation of the spectra difficult. For this reason, it is primarily used as a "fingerprint" that is compared with spectra taken while a process is working well to identify the state or drift of the plasma. As a research and development tool, it can be very useful in understanding the basic processes within the plasma. It can be quite effective and quantitative if combined with other measurements.

A schematic of the optical emission analysis is shown in Fig. 3. Optical emission measurements are usually made through a quartz window on the plasma reactor, with a monochromator system that measures emission wavelengths between 200 and 1000 nm by the rotation of a diffraction grating. A typical scan would take on the order of a minute for this range of frequencies.

Therefore, only one emission energy can be monitored as a function of time during each experimental run in an etching process. Multichannel optical emission analyzers are also available, where diode array detectors are used to collect the spectra from a monochromator with a fixed grating. This technique has the advantage of acquiring a scan in a few seconds, so that multiple scans can be made during a process cycle to characterize transient process behavior.

The required components for setting up an optical emission spectroscopy include:

a) *Optical window*: quartz or sapphire is used to maximize the transmission of short wavelengths of light. Experimental difficulties can be caused by deposition on the chamber window through which the optical emission is sampled. Such deposits can selectively absorb emission, altering the spectra that are observed. In some cases, this can be overcome by purging the window with the input gas to keep the window clean. In addition, heating of the optical windows could also reduce the deposition.

b) *Spectrometers*: Most of the photon detectors have a fairly flat response for different wavelength of the photon, thus, it is essential to disperse the plasma emission into the different wavelengths prior to the photo-detector. Several most commonly used spectrometers are listed here:

- *Prism*: A prism works because the refractive index of glass depends on the wavelength. However, the use of prism in the plasma spectroscopy is not as common these days due to its limited resolving power. Note that the resolving power is defined as $\lambda/\Delta\lambda$ and it defines how well a spectrometer can disperse the light into the different wavelengths.

- *Gratings*: After the invention of the ruling machine by Rowland, the diffraction gratings have been used extensively for optical spectroscopy. Gratings with a resolving power of 100,000 can be easily obtained in a moderate size grating. For example, an 8 cm wide grating with 1200 line per mm has a resolving power of about 100,000 comparable to a glass prism with a base of 80 cm. One of the few shortcomings of the gratings is its relatively small optical throughput when a high resolving power is needed.

- *Filters*: If the wavelength of a particular emission is known and the interference from the nearby wavelength can be neglected, one can use filters to

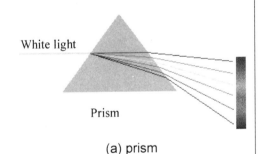

White light

Prism

(a) prism

(b) grating

Fig. 4. Spectrometers

study the emission. A high quality filter can provide about 1 to 2 nm resolution. The main advantages of the filter are its low cost and its high optical throughput, around 80 percent of light transmission.

c) *Detectors*: Photodetectors are essential for collecting the optical emission for the subsequent analysis.

- *Photomultiplier (PM) tube*: The operation principle of a PM tube is based on the photoelectric effect where photons transfer their energy ($h\nu$) to excite the electrons in the materials to a higher energy level, or eject electrons from the surface to a vacuum, as shown in Fig. 5(a). The ejected electron can be accelerated by an external electrical field applied to the cathode, dynodes, and anode. When the accelerated electron hits the dynode surface, it is multiplied and the total number of multiplication would be as many as 10^7 electrons per each photo-ejected electron in less than 10^{-8} second. This large multiplication and the fast time response makes the PM tube one of the most frequently used photon detector.

- *Photodiode (PD)*: Photodiode arrays as shown in Fig. 5(b) are multi-channel light detectors consisting of a few hundreds to thousands of pixels in a linear array. Photoelectrons are generated through closely spaced capillaries coated with low-work function materials to emit secondary electrons and cause a cascade.

- *Charged coupled devices (CCDs)*: CCDs are the product of modern integrated circuits, as shown in Fig. 5(c). In CCDs, the absorbed photon transfers the energy to the electron in the light sensitive detection area and excite the electron into the conduction band. With the externally applied voltage, this excited electron move through the detection area and store in the adjacent capacitor. Accumulation of charge will occur as more electrons are moved to the capacitor until the capacitor is discharged for readout. The advantages of the CCDs include their very small detector dimension, high sensitivity, and very high signal to noise ratio. A typical dimension of a CCD detector is only about 10 to 20 μm, thus a CCD with 1024×1024 array of pixels will be only about 2×2 cm^2. The more sensitive CCDs now approaches 0.8 quantum efficiency (0.8 charge per each photon hitting the CCD). Due to the large number of pixels, however, the readout period is often longer and the time

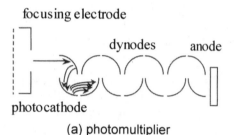

(a) photomultiplier

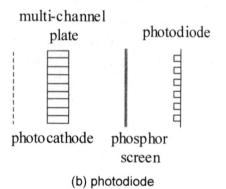

(b) photodiode

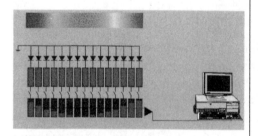

(c) charge-coupled device

Fig. 5. Detectors

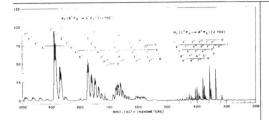

Fig. 6. Optical emission spectrum of a N₂ plasma.

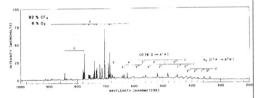

Fig. 7. Optical emission spectrum of a 92% CF₄ – 8% O₂ discharge.

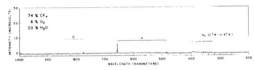

Fig. 8. Optical emission spectrum of a 74% CF₄- 6% O₂ plasma with 20% H₂O.

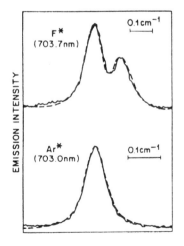

Fig. 9. Line shape determines the collision process.

resolution is usually several KHz.

d) *Other optical components*: There are many different optical components such as lens, mirrors, optical fibers, optical stops, etc., available for various spectroscopic applications. These components are used to enhance the overall performance of the spectroscopic system and should be selected carefully.

Shown in Fig. 6 is a spectrum from a N_2 discharge. A fairly complex spectrum is seen for this fairly simple molecule. The "families" of emission lines are caused by the vibrational states superimposed upon the electronic transitions, for example, N_2 $(B^3\pi_g \rightarrow A^3\Sigma_u^+)$ and $(C^3\pi_u \rightarrow B\ ^3\pi_g)$. If a more complex molecule or mixture is used, the density of lines increases and can cause the observation of what appears to be a continuous band. For example, Fig. 7 and 8 show spectra from a CF_4/O_2 plasma with and without the addition of water vapor. The atomic fluorine emits with very sharp peaks (around 700 nm) that can be easily identified, but CF_3 produces a continuous broad band. The effect of water vapor is clear, since F reacts with H_2O to form HF, F is largely depleted from the plasma and disappeared from the spectrum. If water is the contamination or a leak into the system, it would modify the optical emission spectrum significantly, enabling the detection of system failing with great sensitivity.

The line shape of the optical emission peaks allows the differentiation of various excitation processes, such as differentiating the radiation of atomic species due to direct and dissociative excitation. Typically, the atomic lines are very sharp if the atoms originate directly from the electron impact excitation. The atoms originate from electron impact dissociation of molecular species is more diffusive and broader because dissociative excitation generally results in excited neutral fragments having several volts of energy, the radiation is Doppler broadened and can therefore be distinguished from the much sharper linewidth for radiation produced by direct excitation of a room temperature atom. Subtracting the emission intensity in the broadened tail from the total intensity allows the intensity due to direct excitation alone to be determined.

As shown in Fig. 9, optical emission of F in a $CF_4/Ar/O_2$ plasma is recorded. The line shape of optical emission of F* at 703.7 nm is as sharp as the emission of Ar* at 703.0 nm, indicating that the F originates from direct electron impact excitation of the F atom:

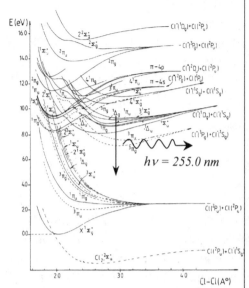

Fig. 10. Energy diagram of Cl$_2$

$3s^2.3p^4(^3P)4p$

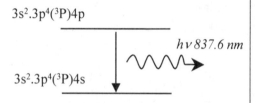

$h\nu\ 837.6\ nm$

$3s^2.3p^4(^3P)4s$

Fig. 11. Partial energy diagram of Cl

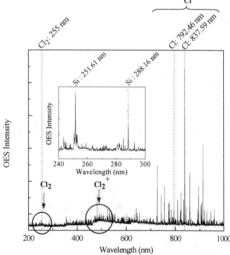

Fig. 12. Optical emission from a Cl$_2$ plasma (Si as the substrate).

$$e^- + F \rightarrow F^* + e^-$$

If the optical emission originates from the dissociation of molecular fluorine:

$$e^- + F_2 \rightarrow F + F^* + e^-$$

much broader optical emission lines will be observed.

Figures 10-12 show the energy diagrams of Cl$_2$ and Cl, and the optical emission spectrum from a chlorine discharge, outlining the signatures of Cl (837.6 nm), Cl$_2$ (255 nm), and Cl$_2^+$. Chlorine is commonly used in the microelectronics industry for etching various materials, and the densities of Cl, Cl$^+$, and Cl$_2^+$ increased monotonically with power at a constant chlorine pressure, whereas the Cl$_2$ densities reduced.

As stated before, the line shapes of the optical emission peaks aid to differentiate the various excitation processes. A set of detailed optical emission spectra from a Cl$_2$/Ar plasma is shown in Fig. 13. The Cl* emission at the anode is sharp, indicating an electron impact excitation of atomic chlorine. The Cl* at the cathode comprises of two components: one sharp feature and one broader peak associated with the dissociative excitation (as shown with the two colored fitted lines).

$$e^- + Cl \rightarrow Cl^* + e^-$$
$$e^- + Cl_2 \rightarrow Cl^* + Cl + e^-$$

A number of different monitors have been developed to indicate the completion of etching. They can be categorized as measuring either completion at a particular point on a wafer (e.g. laser optical reflectivity) or measuring an average value for all the wafers. The common difficulty among all of the end point detectors is discriminating the correct time of completion, since the output is a continuous function that often has only a minor change in derivative as the process is completed. Discrimination is especially difficult if the process is not uniform across the reactor, leading to an even less abrupt change in the process at completion.

3. Actinometry

To interpret the optical emission spectra (OES) of a plasma the kinetics of excitation and relaxation must be considered. The optical system contains emission at a large number of wavelengths, corresponding to allowed transitions between a combination of electronic, vibrational, and rotational states. The electron impact is responsible for creating the excited neutrals, so the electron distribution function must be known in order to calculate quantitatively the concentration of neutrals from

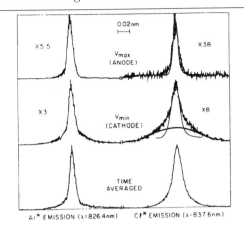

Fig. 13. Line broadening indicates
different processes.

the measured emission intensity.

To illustrate the analysis, let n_A be the concentration of the free radical A and let I_λ be the optical emission intensity integrated over the linewidth. The emission due to excitation from the ground state A can be written as

$$I_\lambda = \alpha_{\lambda A} n_A \qquad (1)$$

with

$$\alpha_{\lambda A} = k_D(\lambda) \int_0^\infty 4\pi v^2 dv Q_{A*} \sigma_{\lambda A}(v) v f_e(v) \qquad (2)$$

Here $f_e(v)$ is the electron distribution function, $\sigma_{\lambda A}$ is the cross section for emission of a photon of wavelength λ due to electron impact excitation of A, Q_{A*} is the quantum yield for photon emission from the excited state ($0 \le Q_{A*} \le 1$), and k_D is the response constant of the photodetector. For low-pressure plasmas and excited states having short lifetimes, $Q_{A*} \approx 1$, though Q_{A*} is generally less than unity for metastable states, due to collisional or electric field de-excitation, ionization, or other processes that depopulate the state without emission of a photon. We note that the cross section $\sigma_{\lambda A}$ differs from the cross section σ_{A*} for excitation of A to level A*, since spontaneous emission to more than one lower lying level can occur. These two cross sections are related by: $\sigma_{\lambda A} = b_\lambda \sigma_{A*}$, where b_λ is the branching ratio for emission of a photon of wavelength λ from the excited state A*.

Typically $\sigma_{\lambda A}$ is known but $f_e(v)$ is not; i.e., $f_e(v)$ is not generally a single-temperature MBD. As plasma operating parameters (pressure, power, driving frequency, reactor size) are varied, $f_e(v)$ changes shape. In particular, the high-energy tail of the distribution, near the excitation energy E_{A*}, can vary strongly as discharge parameters are changed. Consequently $\sigma_{\lambda A}$ changes and $I_\lambda = \alpha_{\lambda A} n_A$ is not proportional to n_A. This limits the usefulness of a measurement of I_λ, which provides only qualitative information on the radical density n_A.

To quantify the concentration of the excited species, Coburn and Chen used a small concentration of inert trace gas of known concentration, n_T, within the plasma (termed an actinometry gas) to interpret more quantitatively the emission spectra and determine the radical density n_A of interest. By comparing the relative intensity of emission from the reference trace gas and the species of unknown concentration, they compensated for variations in the electron distribution function. Typically a noble gas such as Ar is used as it does not react, and therefore its concentration is relatively constant as parameters such as

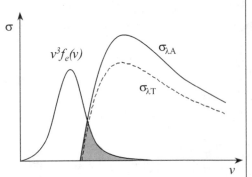

Fig. 14. Principle of actinometry.

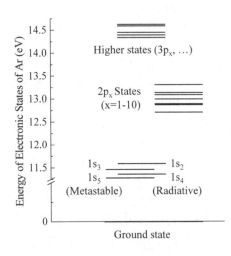

Fig. 15. Energy level diagram of Ar, a common actinometer.

plasma power are varied. The concentration of Ar is computed using the ideal gas law based on the total process pressure and the fraction of gas that is used.

Now choose an excited state T* of the trace gas T that has nearly the same excitation threshold energy, $\varepsilon_{T*} \approx \varepsilon_{A*} \approx \varepsilon_*$. The cross sections $\sigma_{\lambda A}(v)$ and $\sigma_{\lambda'T}(v)$ for photon emission of λ (from A) and λ' (from T) are sketched in Fig. 14. A typical form for the multiplication factor $v^3 f_e(v)$ is also shown as the overlapping shaded area. For the tracer gas,

$$I_{\lambda'} = \alpha_{\lambda'T} n_T \qquad (3)$$

with

$$\alpha_{\lambda'T} = k_D(\lambda') \int_0^\infty 4\pi v^2 dv \, Q_{T*} \cdot \sigma_{\lambda'T}(v) v f_e(v) \qquad (4)$$

Since there is only a small range of overlap of $f_e(v)$ with σ, we approximate cross-sections with values near the threshold: $\sigma_{\lambda'T} \approx C_{\lambda'T}(v - v_{thr})$ and $\sigma_{\lambda A} \approx C_{\lambda A}(v - v_{thr})$, where the C's are the proportionality constants. We then take the ratio of I_λ and $I_{\lambda'}$ to obtain

$$n_A = C_{AT} n_T \frac{I_\lambda}{I_{\lambda'}} \qquad (5)$$

where

$$C_{AT} = f\left(k_D(\lambda), Q_{A*}, C_{\lambda A}, k_D(\lambda'), Q_{T*}, C_{\lambda'T}\right) \qquad (6)$$

It is often possible to choose $\lambda' \approx \lambda$ such that $k_D(\lambda') \approx k_D(\lambda)$, and assumes $Q_{A*} \approx Q_{T*}$. Hence the constant of proportionality is related to the threshold behavior of the two cross sections. If n_T is known and I_λ and $I_{\lambda'}$ are measured by OES, an absolute value of n_A can be determined. Even if C_{AT} is not known, the relative variation of n_A with variation of plasma parameters can be found. A perfect actinometry trace gas would have a cross section for excitation that is identical to that of the species of interest. Any variations in the electron energy distribution would also cancel. However, this is difficult to find. In practice, the electron temperature in low density plasma is fixed by the electron impact ionization and dissociation processes. At lower power densities, a small fraction of the gas is dissociated and little variation is observed with power, pressure, or electrode spacing. Therefore, actinometry has been found to work very well even if the excitation cross sections are not well matched. For F optical emission at $\lambda = 7037$Å with a threshold energy of 14.5 eV, a common choice for the tracer gas is

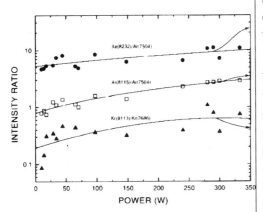

Fig. 16. TRG-OES analysis
[JVST A **15**, 550 (1997)]

argon (as shown in Fig. 15) at $\lambda'=7504$Å with a threshold energy of 13.5 eV. Typically, n_T is chosen to be 1-5% of the feed gas density.

Recently, Malyshev and Donnelly developed the trace rare gases optical emission spectroscopy (TRG-OES) as a new, nonintrusive method for determining electron temperatures (T_e) and estimating electron densities (n_e) in low-temperature, low-pressure plasmas, as shown in Fig. 16. Their method is based on a comparison of atomic emission intensities from equimixture of He, Ne, Ar, Kr, and Xe rare gases to the plasma. For Maxwellian electron energy distribution functions (EEDF), T_e is determined from the best fit of theory to the experimental measurements. For non-Maxwellian EEDFs, T_e is derived from the best fit describes the high-energy tail of the EEDF.

4. Advantages/disadvantages

The primary advantage of optical emission analysis is that it is non-intrusive and can be implemented on an existing apparatus with little or no modification. It provides spatial and temporal resolution of the plasma emission spectra and has very large information content which yields much valuable information about the plasma if analyzed properly. Moreover, it is relatively inexpensive and can be used on more than one reactor.

However, its complex spectrum is often difficult to interpret. Therefore, typically only the atomic lines are used in plasma process analysis. Molecular lines of unknown origin are often used to monitor species whose emission changes significantly upon the end point in plasma etching processes. It is reasonably effective as a trouble-shooting tool to identify contamination as water in the process or an air leak, as long as the "normal" processing spectra haven been recorded. One of the most limiting factors of OES as a process diagnostic tool is the maintenance of the optical window. Deposition and/or etching of the window can significantly modify and attenuate the OES signal.

5. Application: end-point detection

End point detection is used to determine the end of a plasma etching process to better control etching fidelity. The use of single wafer etchers where the process is terminated after a standard over-etch time determined by end point detection has resulted in improvement of reliability and reduced process variation.

The control of the over-etching time is important to the quality of etching processes. It is necessary to have a

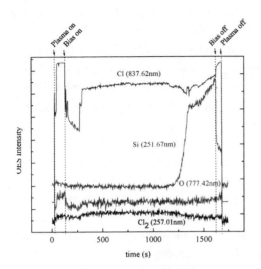

Fig. 17. End point detection in Cl$_2$ etching of ZrO$_2$.

sufficient over-etch to assure that the thin film has cleared at all points, and to remove stringers that result from locally thicker regions where the films are deposited over steps. The excess etchant causes an acceleration of the etching of the remaining horizontal surfaces of the film.

Figure 17 shows the end-point detection of Cl$_2$ plasma etching of ZrO$_2$ on Si by OES: the optical emission from the reactant including atomic chlorine at 837.62nm, molecular chlorine at 257.01nm and atomic oxygen at 777.42 nm and etching products such as atomic silicon at 251.67 nm are monitored for end point detection. As the plasma was turned on, very strong Cl intensity was detected, and a decrease in intensity was observed after applying the substrate bias due to a manual adjustment of the matching networks for both the microwave source power and the RF substrate bias. Once the matching yielded a stable plasma, Cl, Cl$_2$, and O intensities remained relatively constant, until Si intensity was detected upon reaching the etching end-point. Upon the appearance of atomic Si emission, a slight decrease in both Cl and Cl$_2$ intensities was observed. This is likely due to more chlorine consumption in etching silicon, which caused a measured chamber pressure decrease. As the pressure decreases, the electron temperature increases to dissociate Cl$_2$ and excite Cl more effectively, resulting in a continued decrease in Cl$_2$ intensity and a recovery in Cl intensity.

It is clear that optical emission can be used in the detection of endpoints. Alternatively, several other techniques suitable for end-point detection are summarized in Table 2 for comparison.

Table 2. End-point detection techniques

Method	Measuring	Monitoring	End point
Emission Spectroscopy	Intensity of light emitted from discharge	Emission from reactive species and/or etch products	Average for all wafers
Full Wafer Interferometry	Intensity of light interference	Changes in film thickness	One wafer and in situ etch rate monitor
Optical Reflection	Interference phenomena or reflectivity differences	Changes in film thickness	One wafer and in situ etch rate monitor
Mass Spectrometry	Gas composition	Etch products	Average for all wafers

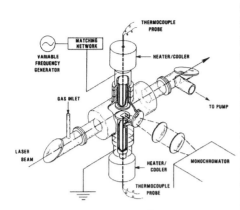

Fig. 18. Laser induced fluorescence setup.

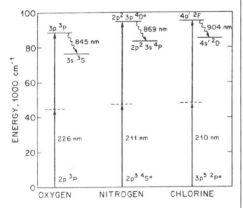

Fig. 19. A two photon excitation process for oxygen, nitrogen, and chlorine.

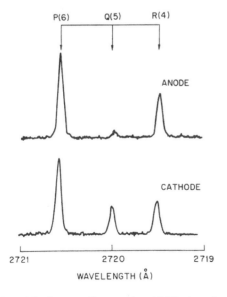

Fig. 20. Spectrally resolved LIF signals of a Cl_2 plasma.

II. LASER INDUCED FLUORESCENCE

Laser induced fluorescence (LIF) is a technique where the emission is created by external light stimulation. The relaxation of the excited states yields photons of specific wavelengths that can be used to identify various species in the plasma. A schematic diagram of the laser induced fluorescence is shown in Fig. 18.

One of the most widely used laser induced fluorescence technique involves a two-photon excitation process, where an excimer laser or a Nd:YAG laser is used to induce two or multiple photon excitation to aid the identification of atomic species such as atomic oxygen. For examples, an incident laser of 226 nm wavelength can induce a two-photon-excited optical transition of atomic oxygen: 2p (^{3}P) $\rightarrow$ 3p (^{3}P). Then the fluorescence occurs at 844nm due to the relaxation transition: 3p (^{3}P) $\rightarrow$ 3s (^{3}S), as shown in Fig. 19.

LIF can also be used to investigate the effect of an electric field on the P, Q, and R- branch of the rotational spectra, as shown in Fig. 20. Though Q- branch is forbidden by the selection rule, the electric field can cause Stark interaction and the Q- branch gains intensity at the expense of the P and R branches. This effect is corroborated by the first principle calculations and proves the capability and sensitivity of such measurement.

These optical techniques can be combined for corroboration and improvement of the ability to quantify the concentration of plasma species. Shown in Fig. 21 are results from an O_2/CF_4 plasma mixed with 2-3% Ar added as a tracer gas. The oxygen concentration was determined by actinometry using O atom emission at two different wavelengths, λ=7774Å (3p^{5}P$\rightarrow$3s^{5}S) and λ=8446Å (3p^{3}P$\rightarrow$3s^{3}S), each ratioed to the Ar emission at wavelength λ'=7504Å. When compared to the two-photon LIF measurement, it can be seen that the 8446/7504 Å actinometric measurement tracks the two photon LIF measurement fairly well as the CF_4% is varied. However, the 7774/7504 Å measurement yields a saturation of oxygen concentration rather than a decrease as the CF_4 concentration is lowered to below 20%, contrary to the LIF measurement.

Again, this is due to the competition of the dissociative excitation process:

$$e^- + O_2 \rightarrow Cl^* + Cl + e^- \rightarrow 2Cl + e^- + h\nu$$

with the direct excitation process:

$$e^- + O \rightarrow O^* + e^- \rightarrow O + e^- + h\nu$$

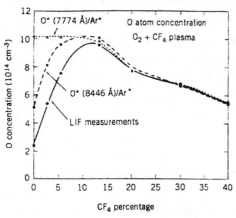

Fig. 21. Comparison of OES and LIF.

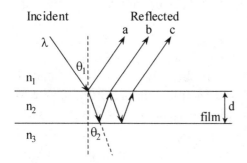

Fig. 22. Schematic drawing of a laser
interferometry apparatus.

such that the measured optical emission intensity

$$I_\lambda = \alpha_{\lambda O} n_O + \alpha_{\lambda O_2} n_{O_2}$$

has a component proportional to the O_2 feed density n_{O_2} as well as the atomic density n_O. The actinometric measurement of n_O will fail if $\alpha_{\lambda O} n_O \leq \alpha_{\lambda O_2} n_{O_2}$, which is the case for the 7774 Å measurement.

III. LASER INTERFEROMETRY

Laser interferometry measures the reflection of a laser from the thin film being etched. At each interface, the light interacts with both media and will be reflected, absorbed, or transmitted. Interference between light reflected from the upper surface and the lower interface of the film causes a periodic variation of the reflected laser beam as the film thickness varies, as shown in Fig. 22. The incident light is at a wavelength of λ, travels through a medium with a refraction index of n_1 (=1 air), and encounters a thin film of thickness d, with a refractive index of n_2 at an angle of θ_1. The incident light reflects from the surface of the thin film to yield a reflected "a" ray; it also reflects off the interface between the thin film and the substrates (index n_3) to yield the "b" ray. The "c" ray is due to a multiple reflection. Due to the phase shift associated with the angle of incidence and film thickness, constructive/destructive interference can occur.

At normal incidence, the peaks and valleys of the reflectivity as a function of time correspond to film thicknesses passing through ¼ wavelength multiples of the laser light in the thin film. This technique can be used on patterned thin films to determine the etching rate of the film, if the film does not cover too large a percentage of the surface.

Application of this technique requires an abrupt change in refractive index at the thin film/substrate interface. A sufficiently thin and transparent film is required to ensure some laser reflection at the interface. The interference created between the light reflected from the film being etched and the underlying thin film produces a sinusoidal intensity that can be used to monitor the etching rate of the thin film. For a laser beam striking the film at an angle θ_1 to the normal, the reflected beam amplitude oscillates due to interference, and the change of film thickness d is given by:

$$\Delta d = \frac{\lambda / 2}{\sqrt{n_2^2 - \sin^2 \theta_1}} \qquad (7)$$

Laser interferometry measures the periodic

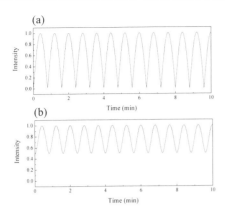

Fig. 23. Computed interferometry signal without photoresist. The interference pattern (a) originates from a more reflective thin film than that in (b).

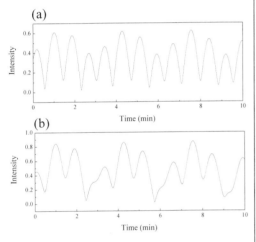

Fig. 24. Computed interferometry signal with equally-lined-and-spaced photo-resist. The interference pattern (a) originates from a less reflective photoresist.

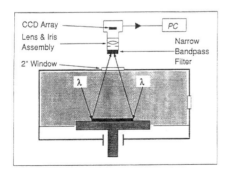

Fig. 25. Setup for performing the full wafer interferometry.

reflection from a film in which the extinction coefficient of the laser light is sufficiently low, e.g., polysilicon and oxide can be readily measured. For metals and most silicide films, no periodicity is seen because of their high extinction coefficients (the light is adsorbed in the film), but an abrupt change in reflectance is recorded upon clearing.

Using etching of Si on SiO_2 as an example, the interferometry pattern can be easily determined by specifying the index of refraction of all materials and the reflectance, transmittance, and absorbance of all materials. Figure 23 shows the computed interference pattern of blanket Si on SiO_2: (a) 50% reflection and 50% transmission for Si and 100% reflection for SiO_2, and (b) 25% reflection and 75% transmission for Si and 100% reflection for SiO_2. The variation in the interferometric amplitude is larger in the case where the thin film is more reflective.

Figure 24 shows the computed interference pattern of equally-lined-and-spaced photoresist pattern on Si on SiO_2: (a) 25% reflection and 75% absorption for photoresist, 50% reflection and 50% transmission for Si and 100% reflection for SiO_2, and (b) 75% reflection and 25% absorption for photoresist, 50% reflection and 50% transmission for Si and 100% reflection for SiO_2. Due to the presence of the photoresist pattern, the interferometric signal was modulated at periods greater than the interferometric signal in the absence of the photorsist patterns. This modulation is more significant when the reflectivity of the photoresist is larger.

The laser interferometry is quite inexpensive, but may require modification of the plasma reactor if it is not equipped with the appropriate windows. Minor difficulties can occur if the window becomes dirty. The technique provides a continuous measure of the etching rate, displays any inhibition period, monitors the variation in etch rates, and detect the onset of surface roughening by a continuous decrease in amplitude of the reflectance.

IV. FULL-WAFER INTERFEROMETRY

Full wafer interferometry uses a CCD camera interface with a computer to collect the modulation of reflectivity from the wafer as a function of time, and thereby, determine the etching rate of thin film materials or endpoint of highly reflective materials, as shown in Fig. 25.

Full wafer interferometry is able to measure etching uniformity across a wafer, within dies, and aspect ratio

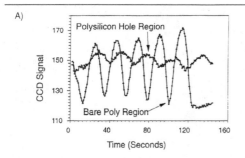

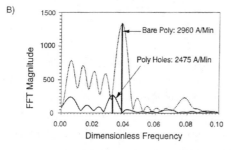

Fig. 26. Interferometry signal for polysilicon measured by the CCD technique.

dependent etching, if structural information within the die is available. This technique uses the light emission from the plasma unlike laser interferometry where a laser is used to illuminate the wafer. The resolution of etching rate is approximately 1 %. It also works for submicron features as the interference occurs in the unetched region, as long as the wavelength of the interfering light is longer than the feature size. For example, measurements of 0.5 μm patterned lines and spaces have been performed with 0.6 μm light.

As shown in Fig. 26 (a) during the etching of patterned polysilicon, the actual CCD signal measured for both a bare polysilicon region and a region patterned with photoresist. The patterned polysilicon region shows an attenuated signal due to the presence of the photoresist. Figure 26 (b) presents the magnitude of the FFT for the signals in Fig. 26 (a). The dotted line is the FFT spectrum for the bare polysilicon region, which yields an etching rate of 296 nm/minute. The solid line is the FFT spectrum for the patterned polysilicon region, which yields a lower etching rate of 248 nm/minute.

V. MASS SPECTROMETRY

Mass spectrometer has long been used to measure the composition of a gas by ionizing the gas, typically by electron impact ionization, and separating the resulting ions by the mass-to-charge ratios, and the set up is typically as shown in Fig. 27. For a time-of-flight mass analyzer with a flight tube of L, the ions are dispersed in time:

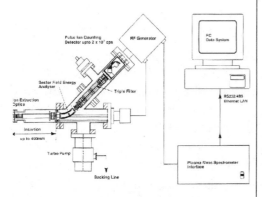

Fig. 27. Schematic drawing of a mass spectrometer setup.

$$E = mv^2 / 2 = zV$$
$$v = \sqrt{2E / m} \qquad (8)$$
$$t = L / v = L\sqrt{m / 2Vz}$$

The field created by the quadrupole structure causes the ions to transverse between the poles in a helical path that is stable for only a particular mass-to-charge ratio (Fig. 28). The ions that successfully pass through the mass filter are collected and the resulting current is amplified to form the mass spectrometer signal. Varying the AC and DC potentials allows the selection or scanning of mass-to-charge ratios, as shown in Fig. 29.

To avoid detecting species that are modified by collisions with chamber walls, the mass spectrometer must be mounted as close to the discharge as possible, so that the gas entering the mass spectrometer chamber has minimal opportunity for contacting the walls. In most plasma processes, the mass spectrometer measures the gas

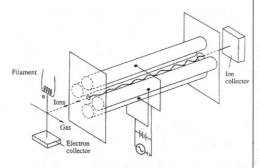

Fig. 28. Schematic drawing of a quadrupole mass spectrometer.

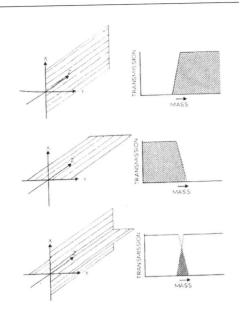

Fig. 29. Operation of a quadrupole mass spectrometer. The alternating RF field results in a narrow mass peak of a certain mass-to-charge ratio.

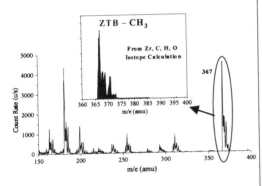

Fig. 30. Fragmentation pattern of zirconium t-butoxide (ZTB).

composition at a point between the chamber and the pumping system. However, for research studies, the mass spectrometer can be built into the system so that it measures the ion and neutral flux incident upon an electrode through the use of an orifice in the electrode. With this technique, both the ion and neutral fluxes upon the electrode can be recorded, and the species to be detected pass from the plasma reactor into the ionizer of the mass spectrometer in a line-of-sight path, minimizing the corruption of the gas composition caused by collision of gas with walls.

In a plasma, the neutrals, radicals, and ions in the plasma can be sampled through an orifice for analysis. Ions can be directed analyzed based on their mass-to-charge ratio. For detecting neutrals and radicals, ions need to be deflected and neutrals be ionized for detection. During electron impact ionization, the fragments created from a number of different species could have the same mass-to-charge ratio, thus a quantitative analysis requires the measurement of the cracking patterns for each species, and a comparison of these individual spectra to the composite spectrum. From this, the combination and quantity of species whose spectra would sum to the composite spectrum can be determined.

As an example, the PECVD of ZrO_2 using an organometallic $Zr(OC_4H_9)_4$ precursor and O_2 is discussed here. The cracking pattern of $Zr(OC_4H_9)_4$ is first characterized in the absence of plasma, as shown in Fig. 30. Several Zr-containing fragments could be easily identified in the mass spectrum using the relative abundance of naturally occurred Zr isotopes, i.e., ^{90}Zr (51.45 %), ^{91}Zr (11.22 %), ^{92}Zr (17.15 %), ^{94}Zr (17.38 %), and ^{96}Zr (2.80 %). The highest mass and most abundant fragment corresponds to a ZTB molecule missing one methyl group, $Zr(OC_4H_9)_3(OC_3H_6)$, at m/z=367-373, whose isotopic mass patterns exactly matched those calculated based on the relative elemental isotopic abundances, as shown in the inset of Fig. 30. With the cracking pattern determined, the quantitative analysis is thus possible.

In Fig. 31, three ion mass spectra in the m/z range of 85-170 from the ZTB plasmas with O_2/ZTB(Ar) ratios of 0, 0.5, and 4 are shown at the same pressure and microwave power. The ion intensities were normalized to the Ar flow rate, which scales with the feed rate of ZTB precursor, and the normalized intensity of the most abundant $ZrO_xH_y^+$ in each spectrum was comparable. It is clear that Zr^+ and ZrO^+ are the dominant Zr-containing

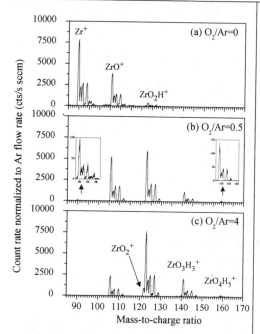

Fig. 31. Relative abundance of ZrO$_x$ as a function of O$_2$ addition to ZTB.

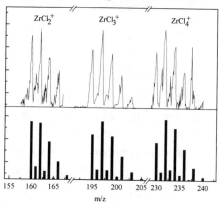

Fig. 32. Etching products from Cl$_2$ etching of ZrO$_2$.

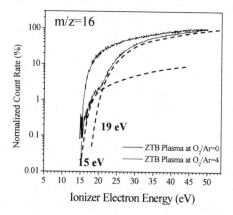

Fig. 33. Appearance potential mass spectrometry allows the differentiation of dissociative ionization products.

ionic species in the plasma with no O$_2$ addition (O$_2$/Ar =0). Even ZrO$_2$H$^+$ existed in a small amount, though Zr-containing ions with higher oxidation states were not detected. This result suggests, first, both Zr-O (yielding Zr$^+$) and C-O dissociations (yielding ZrO$^+$ and ZrO$_2$H$^+$) in the ZTB precursor molecule take place in the plasma gas phase and their probabilities are comparable. Secondly, Zr$^+$, ZrO$^+$, and ZrO$_2$H$^+$ were produced through multiple sequential electron collisions, since ZrO$_3$H$_3^+$ and ZrO$_2$H$^+$ were the dominant species observed in the neutral mass spectrum in the absence of a plasma, where most fragments were derived through a unimolecular decomposition of the ZTB precursor. The following sequential dissociation reactions are thus suggested to occur in the plasma:

$$\cdots \rightarrow ZrO_3H_3 \xrightarrow{e^-} ZrO_2H \xrightarrow{e^-} ZrO \xrightarrow{e^-} Zr$$
$$\searrow \qquad \searrow \qquad \searrow \qquad (9)$$
$$O, 2H \qquad O, H \qquad O$$

Similarly, mass spectrometry can be used in detecting plasma etching products, especially polyatomic species, which are not sensitive to the OES analysis. Figure 32 shows the etching products in Cl$_2$ etching of ZrO$_2$ thin film, where ZrCl$_x$ are the dominant etching products.

Finally, appearance potential mass spectrometry allows the detection of a wide range of radicals and identify their origin from different excitation processes. In this method, the ionization filament energy is gradually increased while the species of interest at a particular mass-to-charge ratio is monitored. The appearance of the chemical species at a particular electron energy of the ionizer would allow one to discriminate the radicals from different origins. For example, Fig. 33 shows the appearance potential mass scan at m/z=16, which corresponds to O, in a ZTB/O$_2$ plasma. The ionization potential of O is 13.6 eV, but there is no mass signal detected at 13.6 eV, suggesting that the detected oxygen atoms are all from the dissociation processes. There are at least two origins of O. Knowing that the O-O bond strength is about 5.1 eV, the mass signal detected at 19 eV is likely due to O from dissociation of O$_2$, and the mass signal detected at 15 eV is likely to correspond to O from dissociation of C-O or Zr-O in the plasma. This is similar to using the line shapes in the optical emission spectra to differentiate the excitation processes from which the species of interest originate.

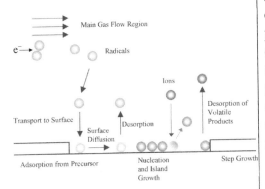

Fig. 1. Schematic diagram of plasma surface interaction.

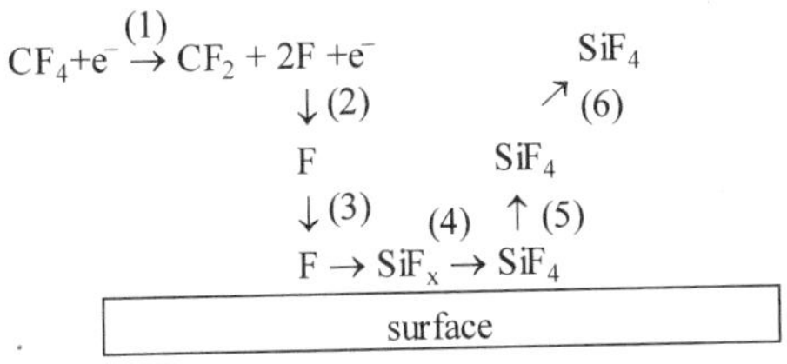

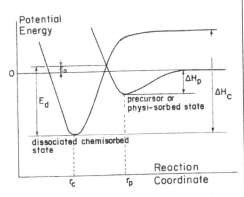

Fig. 2. A descriptive schematic of plasma gas-phase and surface reactions.

Fig. 3. Potential energy diagram for physisorption and chemisorption.

PRINCIPLES OF PLASMA PROCESSING
Course Notes: Prof. J. P. Chang

PART B6: PLASMA SURFACE KINETICS

I. PLASMA CHEMISTRY

To characterize plasma-surface interactions, we first reviewed elementary reactions, gas phase kinetics, and surface kinetics. Plasma processing shares a relatively common set of steps, by which the surface reaction proceeds. Ion bombardment can alter the kinetics of one or more of the steps, creating an enhancement of the etching or deposition rate; this effect is thought to be the primary cause of anisotropy in the surface topographical change. A reasonable set of steps for plasma etching that can be used to understand the etching mechanisms are as follows (Figs. 1 and 2):

1. Creation of the reactive species within the plasma phase by electron-neutral collisions and subsequent chemical reactions:

$$e^- + CF_4 \rightarrow CF_3 + F + e^-$$

2. Transport of the reactive species from the plasma to the substrate.

3. Adsorption of the reactive species on the surface (either physisorption or chemisorption, Fig. 3).

4. Dissociation of the reactant, formation of chemical bonds to the surface, and/or diffusion into the substrate with subsequent formation of the desorbing species:

$$F^* + SiF_x \rightarrow SiF_{x+1}$$

5. Desorption of the product species:

$$SiF_{4(s)} \rightarrow SiF_{4(g)}$$

6. Transport of the product species into the plasma.

7. Simultaneous re-deposition of etching products.

II. SURFACE REACTIONS

1. Spontaneous surface etching

A number of gas-surface systems of interest to microelectronic fabrication react spontaneously, e.g., F with Si, and Cl_2 with Al. Spontaneous etching is a process in which neutral species interact with a solid surface to form volatile products in the absence of energetic radiation (*e.g.*, ion bombardment or UV radiation). These spontaneous chemical reactions generally are activated and follow an Arrhenius relationship, and the rate of reaction is given by

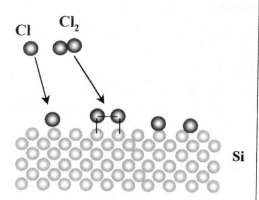

Fig. 4. Spontaneous surface etching reactions.

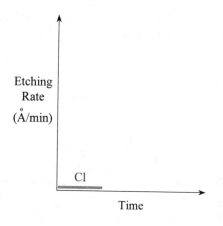

Fig. 5. Slow etching rate of silicon due to Cl spontaneous etching.

$$ER_s = k_o e^{\left(-\frac{E_a}{KT}\right)} Q \qquad (1)$$

where Q is the flux of reactive species, T is the substrate temperature, k_o is the preexponential factor and E_a is the activation energy. The preexponential factors and activation energies for both Cl and F atoms etching of silicon are shown in Table 1 for comparison.

Table 1. Arrhenius rate parameters for Cl and F atoms etching of silicon.

Neutral	Substrate	Q (#/cm^2/s)	k_o (Åcm^{2}s/#min)	E_a {eV}
Cl	Poly-Si	6×10^{19}	2.57×10^{-14}	0.29
F	Si<100>	2.3×10^{19}-1.1×10^{22}	3.59×10^{-15}	0.108

The activation energy of atomic chlorine etching polysilicon is approximately three times larger than that of atomic fluorine. Therefore, the etching yield by atomic chlorine is two orders of magnitude lower than that of atomic fluorine even with a larger preexponential factor. This is consistent with the high energy barriers for penetration of chlorine (13 eV) into the silicon backbones than that of fluorine (1 eV). In the case of silicon etching in chlorine, Equation (1) predicts an etching yield that is 2-3 orders of magnitude less than the overall ion-enhanced etching yield and thus can be ignored (Figs. 4 and 5). However, for Al etching in chlorine, the spontaneous etching is significant.

For a process that is limited by the surface reaction kinetics, the rate is typically a strong function of the surface temperature, however, an etching process that is limited by electron impact reaction in the plasma phase or ion bombardment-induced surface kinetics is relatively insensitive to temperature. An etching process that is limited by a surface chemical reaction produces isotropic etching, since the reactant gas has no strong preferential directionality.

Any free radicals formed will most likely strongly adsorb to the surface, and thus participate in the etching reaction. The creation of the free radical in the gas phase eliminates the chemical barrier for chemisorption that would normally exist at room temperature. The chemical reactions that take place on the surface typically follow a Langmuir-Hinschelwood mechanism, i.e., a reaction between chemisorbed species.

It is worth noting here that the doping level of the

silicon substrate can greatly change the spontaneous etching rate of polysilicon by F and Cl atoms. Houle studied the etching of Si in the presence of F and showed that the doping effect in which heavily n-type polysilicon was etched more rapidly than p-type or undoped, was a result of the band bending at the surface. The adsorbed atomic F began negatively ionized by an electron tunneling from the bulk Si. If the field caused by the band bending aided the transport of F^- into the surface, the reaction rate is accelerated. As the band bending is a function of the Fermi level, the doping changes the etching rate.

In Cl etching of Si, the same doping effect occurs, however, it is more pronounced. Atomic chlorine does not appreciably etch p-type and undoped polysilicon at room temperature, however, it etches n^+-type polysilicon spontaneously with one to two order magnitude increase in etching rate. Ogryzlo termed this effect "field enhanced etching", in that the large electron density in the valence band causes the Fermi level to bend upwards. This band bending facilitates "charge transfer" from silicon lattice to the electronegative and chemisorbed Cl atoms, makes the Si-Cl bonding more ionic, allows for more flexibility in the bonding geometry, and creates more chemisorption sites. The incorporation of chlorine atoms is thus enhanced, as well as the etching rate. The slower etching of Si by Cl and Br than F is probably due to the larger size of Cl and Br and the greater sterric hindrance effect.

2. Spontaneous deposition

Deposition of thin films due to reactive radicals with low volatility is common in both etching and deposition processes. In these reactions, the sticking coefficients of the free radicals are of critical importance of deposition kinetics. For example, SiO_2 deposition by PECVD is widely used as the interlayer isolation between metal lines in MOSFET device, and SiO_2 deposition can be done with different precursor chemistries. Figure 6 shows the conformal deposition of SiO_2 over high aspect ratio features in a TEOS/O_2 plasma, due to the small sticking coefficients of the reactants. Figure 7 shows a non-conformal deposition profile of SiO_2 deposited in a silane/O_2 plasma, due to the large sticking coefficient of the reactants.

In etching processes, deposition of reactants or redeposition of etching products can be useful or detrimental, depending upon the process of interest. In the case of etching of SiO_2, a polymer layer formation is

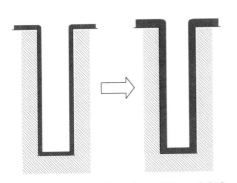

Fig. 6. Conformation deposition of SiO_2 in $Si(OC_2H_5)_4/O_2$.

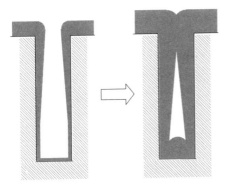

Fig. 7. Non-conformal deposition of SiO_2 in SiH_4/O_2.

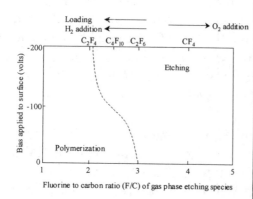

Fig. 8. Plot of boundary between polymer growth and etching (Coburn).

thought necessary for the etching, though too thick a polymer layer would result in an "etch-stop".

Coburn introduced the concept of the carbon/fluorine ratio to help quantify the conditions under which polymer formation occurs. Shown in Fig. 8 is a diagram he developed that characterizes the observations of the effect of the C/F ratio, ion bombardment energy, loading, and additions of H_2 or O_2. As can be seen in Fig. 8, if the feed gas has a high C/F ratio, a polymer can be formed on the surfaces in contact with the plasma.

Polymerization occurs by the sequential addition of free radicals onto a polymer chain. Radicals such as CF and CF_2 can add to a chain without reducing the probability that additional radicals can be added. If CF_3 or F is added to a chain, however, the positions available for additional chain growth are reduced. Therefore, the relative concentrations of these species dictate the growth rate and chain length of polymeric chains. Increasing the F/C ratio favors the formation of F and CF_3, and therefore reduces the propensity for a polymer to be formed in a plasma. Addition of H_2 reduces the concentration of F by reaction to form HF, thereby increasing the effective C/F ratio. O_2 additions react with the carbon, which results in a decreased effective C/F ratio. Increasing the power level to a plasma favors the production of CF and CF_2 over CF_3, but increases the concentration of F sufficiently that the net result is the reduction of polymerization. Increasing the power also increases the probability that a polymer chain is broken by electron impact in the plasma or sputtered from a surface by an ion. Polymer build-up on a surface is decreased by ion bombardment because of sputtering and chain breakage. Therefore, a higher propensity for formation exists on the sidewalls of an etching feature where the ion flux is reduced and polymer can be deposited preferentially.

The addition of oxygen to prevent polymer deposition can also come from the material being etched, i.e., SiO_2. The selectivity of etching with respect to oxide is achieved by balancing the plasma chemistry such that polymer formation does not occur on an oxide surface as it is etched since the oxide supplies enough oxygen which reacts with the carbon to prevent its build up. For the same chemistry, when Si or Al is encountered, a polymeric film builds up on the surface and the etching stops. This means of selectivity allows for very high selectivity for oxide etching with respect to underlying semiconductor or metal films.

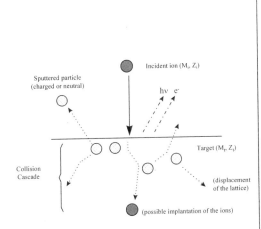

Fig. 9. Surface kinetics during sputtering.

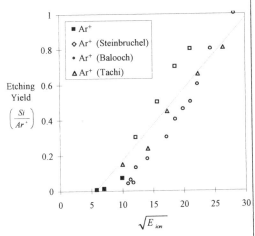

Fig. 10. Physical sputtering yield of polysilicon by Ar^+.

3. Ion sputtering kinetics

The characterization of the physical sputtering yield at the low energy regime was attempted by employing the collision cascade model first proposed by Sigmund to characterize the sputtering yield of amorphous and polycrystalline targets. The sputtering yield can be approximated by assuming that ions slow down randomly in an infinite medium, considering elastic scattering (nuclear stopping), inelastic scattering (electron scattering), and surface binding energy. A schematic diagram of a collision cascade is shown in Fig. 9.

In the low ion energy regime (< 1 kV), the binary particle interactions can be characterized by a Born-Mayer-type cross section, and the sputtering yield is linear to the square root of ion incident energy and can be described as following:

$$Y(E_{ion}) = C_{it} \cdot S_n\left(\frac{E_{ion}}{E_{it}}\right) \qquad (2)$$

C_{it} and E_{it} are constants dependent on the particular ion – target combination (Z_i and Z_t are atomic numbers, and M_i and M_t are masses of the ion and target atoms, respectively), and $S_n\left(\frac{E_{ion}}{E_{it}}\right)$ represents the nuclear stopping power. However, this model overestimates the etching yield by at least a factor of 5 at energies lower than 75 eV, where the ion incident energy is on the same order of magnitude of the surface binding energy. Therefore, the threshold energy has to be taken into account to properly model the sputtering yield at low ion energies.

The universal energy dependence of ion bombardment induced etching processes proposed by Steinbrüchel is therefore used to model most of the experimental measurements. An empirical form was proposed as follows:

$$Y(E_{ion}) = C_{it} \cdot S_n\left(\frac{E_{ion}}{E_{it}}\right) \cdot f\left(\frac{E_{th}}{E_{ion}}\right) \qquad (3)$$

where the modified nuclear stopping function was further modified by Wilson and $f(E_{th}/E_{ion})$ expanded by Matsunami, thus the etching yield can be expressed as:

$$S_n\left(\frac{E_{ion}}{E_{it}}\right) \propto E_{ion}^{1/2}, \text{ and } f\left(\frac{E_{ion}}{E_{it}}\right) = 1 - \left(\frac{E_{th}}{E_{ion}}\right)^{1/2} \qquad (4)$$

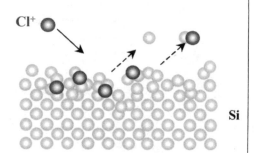

Fig. 11. Sputtering yield of polysilicon by Cl⁺ in the low energy regime, in comparison to molecular dynamic simulation results and low energy sputtering yield by Ar⁺.

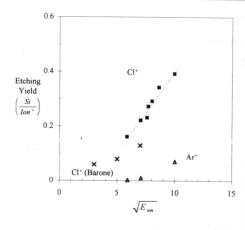

Fig. 12. Chemical sputtering during etching.

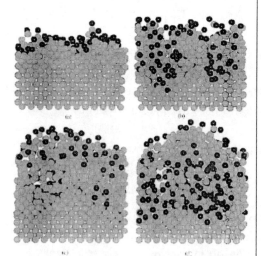

Fig. 13. Molecular dynamic simulation of Cl⁺ interacting with Si [Barone and Graves].

$$Y(E_{ion}) = A \cdot (E_{ion}^{1/2} - E_{th}^{1/2}) \qquad (5)$$

where E_{th} is the threshold energy and A is a constant depending on the particular ion-target combination. Good agreement is observed by employing this universal energy dependence of ion bombardment induced etching processes to Ar⁺ etching of polysilicon, as shown in Fig. 10.

In the case of reactive ion sputtering (chemical sputtering), reactive ions yield higher etching rate than that of inert ions due to the formation of volatile species with the reactive etchant. The sputtering yield of polysilicon by Cl⁺ is a linear function of the square root of the ion energy, as shown in Fig. 11, and is much higher than that by Ar⁺ ions. The extrapolated threshold energy is approximately 10 eV, which is lower than that measured by Ar⁺ sputtering of polysilicon (~35 eV). The reduction in threshold energy is caused by the formation of a heavily chlorinated layer that reduces the surface bonding energy and allows for subsequent incorporation of chlorine into the silicon lattice (Fig. 12). Good agreement is observed by comparing the experimentally measured sputtering yield to molecular dynamic simulation results reported by Barone and Graves (Fig. 13). They confirmed the square root ion energy dependence of reactive chlorine ion etching of silicon over the energy range of 10 to 50 eV, and observed that the etching product stoichiometry depends strongly on the ion energy. Therefore, in comparison with the spontaneous etching, higher etching rate is achieved, as shown in Fig. 14.

In a sputtering process, most of the flux is ejected by momentum transfer, in which the ion bombarding the surface collides with individual surface atoms transferring momentum with each collision. The energetic collision partners are displaced and also undergo a series of binary collisions over a time period of about 10^{-12} seconds.

Etching results when an atom at the surface receives enough energy that it can break loose from the surface potential and leave the surface. For a normal incident ion, at least two collisions are needed for the momentum's direction to be reversed and a surface atom to be sputtered. For ion bombardment at an angle of 40° or greater off normal, only one collision is needed for a surface atom to be dislodged, i.e., sputtered.

As the ion incident angle exceeds 70°, the sputtering yield drops off significantly because the ion tends to reflect from the surface dissipating less energy in the

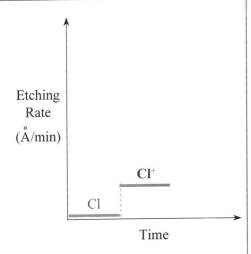

Fig. 14. Higher etching rate achieved by reactive ions compared to reactive neutrals.

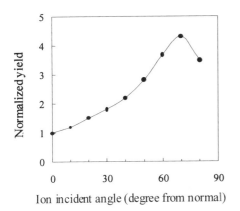

Fig. 15. Dependence of etching yield on ion incident angles.

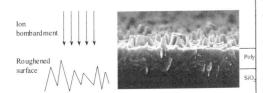

Fig. 16. Formation of "grass" due to micromasking.

collision cascade. Note that the sputtering yield is defined as the number of surface atoms removed per incident ion (Fig. 15). As momentum scales with the square root of energy, the yield also scales with the square root of the ion bombardment energy. It also increases with the mass of the ion. For Ar^+, the sputtering yield of Si is 0.5 at 1 keV and peaks at 1.0 at 3 keV. As the ion energy is further increased, the cross section for scattering at the surface decreases and the sputtering drops. An ion with a greater mass dissipates a greater fraction of its momentum near the surface, increasing the probability that sputtering occurs. There is generally a threshold energy for sputtering which is a few bond strengths associated with the minimum energy it takes an ion to displace a surface atom and thus can be well correlated with other measurements of surface removal energy such as sublimation energy. A typical threshold energy is about 25 eV.

Because of the angular dependence with the largest etch yield at approximately 60°, surface features with about that angle etch more rapidly, resulting in surface facets with a specific angle. An ion-bombarded surface will roughen forming surface peaks or needles that point toward the ion bombardment angle. Their conformation is due to this preferential sputtering angle and redeposition of the sputtered products. If any impurity is present that has a lower etching rate, the impurity will congregate at the point of the needles. In etching process with a large physical sputtering component, the surfaces roughen as they etch (formation of grass as shown in Fig. 16), in particular if carbon, oxide, or some other involatile material is present.

Three other mechanisms occur in sputtering: thermal spikes, surface damage, and electronic excitation. In a thermal spike, the energy of the ion is dissipated as heat, resulting in a short-lived (10^{-8} sec), high-temperature transient located within a short distance of the ion impact. This high temperature induces surface molecules and atoms to desorb rapidly with a kinetic energy that are in equilibrium with the thermal spike temperature. This mechanism is probably not significant for the sputtering of low vapor pressure materials like metals, but could be significant in the sputtering of metal compounds that have higher vapor pressures.

Electronic excitation occurs due to the tunneling of an electron to neutralize the ion just before it strikes the surface. The excited surface state (that has the energy of ionization of the incident ion) can relax by the emission of atoms in some materials. Ions therefore have a slightly greater sputtering yield than neutrals, but this difference can only be observed near the threshold-sputtering yield.

The third sputtering mechanism is the creation of surface damage that degrades into molecular species after the collisional event. These molecular species can then desorb in time and leave with an energy that is characteristic of the bulk temperature of the sample. This can be thought of as an ion-induced degradation of the surface. When a surface is exposed to a flux of reactive neutrals and ions, it has been shown that an adsorbed layer of reactants can form, resulting in a halide-like surface. Thus, ion-induced plasma etching may contain similar mechanistic steps.

Sputtering of a compound results in a shift in stoichiometry of the surface, i.e., the surface becomes richer in the less volatile material. For example, ion bombardment of SiO_2 produces a surface that is rich in Si as the O is more volatile. This enriched layer is thin being of order of a few monolayers in thickness and increases in thickness with the ion bombardment energy.

Knock-on or atomic mixing also takes place during ion bombardment. Ions striking surface atoms can drive those atoms into the lattice several atomic monolayers in depth. This process is termed knock-on and results in the mixing of layers as sputtering takes place.

4. Ion-enhanced chemical etching

When a surface is exposed to both chemically reactive neutral which can react with a surface to produce a volatile product and ion bombardment the combined ion and neutral fluxes often etches more rapidly than surfaces exposed to only the neutral bombardment. Shown in Fig. 17 is a seminal beam experiment performed by Coburn in which the etching rate was monitored while XeF_2 and Ar^+ beams were turned on and off. It can be seen that the combination of both ions and a fluorine source results in an etching rate that is quite synergistic and exhibits an etching yield that is an order of magnitude greater than physical sputtering.

Ion enhanced etching of surfaces by Ar^+, F, and CF_2 has been measured as a function of reactant fluxes as shown in Figs. 18 and 19. It was shown to be a function of the flux ratios, not the absolute values of the flux.

The presence of carbonaceous precursor has been shown to inhibit etching in Si as expected and to enhance the etching of SiO_2. However, at high F fluxes the CF_2 is removed by recombination with F to form CF_4 rather than

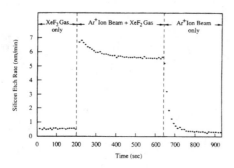

Fig. 17. Etching rate as a function of various beams used in etching (Coburn).

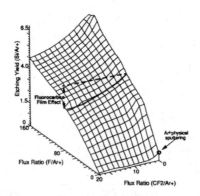

Fig. 18. Ion enhanced etching of Si in Ar^+, F, and CF_2 beams (Gray).

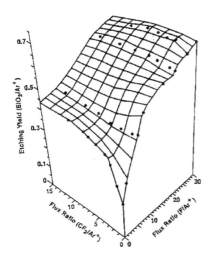

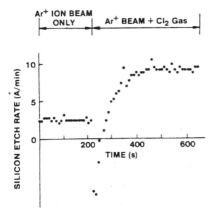

Fig. 19. Ion enhanced etching of SiO₂ in Ar⁺, F, and CF₂ beams (Gray).

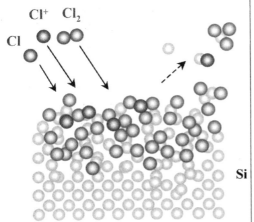

Fig.20. Etching of silicon with Ar⁺ and Cl₂ (Coburn).

Fig. 21. Ion-enhanced etching process.

facilitating the etching of oxide (Fig. 19).

Other studies have demonstrated that greater mass ion that dissipate their energy nearer the surface have a greater ion induced etching yield and that the use of atomic species increases the ion enhanced chemistry.

The energy dependence of etching has been shown to vary with the square root of the ion bombardment energy indicating that the initiation of the etching process is a function of the ion momentum as in physical sputtering. Many researchers, based on this observation suggested that the increased etching yield is due to a reduced binding energy of the surface species, i.e. the chemical reaction with the neutrals leaves the surface atoms more loosely bound. However, the surface residence time for the emission of the products has been measured to be about 10^{-4} seconds, 8 orders of magnitude too large for a physically dominated sputtering process. While physically enhanced etching does occur, the dominant mechanism appears to be the subsequent chemical reactions that occur after the collision cascade.

These beam studies are even more significant enhancement in the etching of Si with Cl₂ for which spontaneous etching does not occur at room temperature. A coverage of approximately a monolayer of Cl is chemisorbed on the surface in the absence of ion bombardment. The quartz crystal microbalance data shown in Fig. 20 shows an initial weight gain and then subsequent etching. The weight gain indicates that Cl is incorporated into the surface by ion bombardment increasing the Cl content. The ion induced etching rate increases with the Cl content. From this data, the incorporation of several monolayer equivalents into the surface at steady state is indicated.

To simulate more realistically the chlorine plasma, the ion-enhanced silicon etching yield was characterized as a function of Cl atom to Cl⁺ ion flux ratio at three ion energy levels. This process and its effect are schematically illustrated in Figs. 21 and 22, while the experimental results are shown in Figs. 23 and 24. In Fig. 23, an initial sharp rise in the etching yield was observed at low flux ratios where reaction was limited by the supply of reactive neutrals. The etching yield then gradually saturated as the reaction became ion-flux limited at high flux ratios. The etching yield was a function of the square root of ion energy. The dotted lines are fits from a surface kinetics model detailed in Eq. (6), as shown later.

In Fig. 24, the effect of ion incident angle shows no significant change as the ion incident angle increased

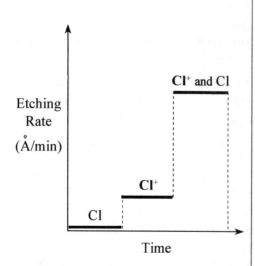

Fig. 22. Ion-enhanced etching increased the etching rate by order(s) of magnitude.

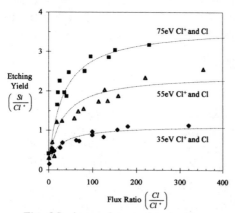

Fig. 23. Ion enhanced polysilicon etching by Cl atoms with Cl⁺ ions as a function of ion energy and neutral to ion flux ratio.

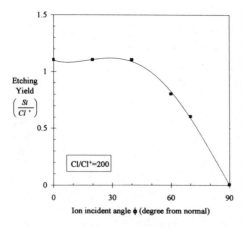

Fig. 24. Angular dependence of the etching yield for ion enhanced etching of Si with combined Cl and Cl⁺ fluxes.

from normal to 40° off-normal, but decreased by 30 % and 50 % at 60° and 70° off-normal, respectively. This angular dependence of ion-enhanced polysilicon etching can be incorporated into a profile simulator to simulate the etching of the sidewalls. It can be seen that unlike the physical sputtering curve shown also, the chemical enhance etching yield does not vary with ion bombardment angle until greater than 40°. This insensitivity to angle allows the etching of features with smooth bottoms. The more rapid drop off with greater angles is thought to be due to ion removal of the adsorbed Cl reducing the chlorination of the surface, and thus, etching rates.

The reduction of etching rates caused by deposition/redeposition of the etching products and by-products was explored by adding a $SiCl_2$ beam -- $SiCl_2$ is produced by etching or by electron impact dissociation of $SiCl_4$ in the plasma. It is known that the concentration of etching products can build up to appreciable levels (~10% of the Cl flux which are inversely proportional to the flow rate). This approach permits a thorough understanding of the fundamental reaction mechanisms and allows formulation of a kinetic model useable in a profile simulator to model the profile evolution during plasma etching processes.

The sticking coefficient of $SiCl_2$ to form a stable $SiCl_x$ film was calculated to be approximately 0.3 based on the measurement of the incident dichlorosilane beam flux and the $SiCl_2$ deposition rate observed by laser interferometry using the index of refraction of polysilicon. As the flux of $SiCl_2$ has been estimated to be on the order of 10% of the Cl flux in high density and low-pressure polysilicon etching processes, use of this sticking coefficient would suggest deposition rather than etching of polysilicon. Figure 25 shows the effect of $SiCl_2$ on chlorine ion-enhanced etching yield in the presence of Cl⁺ and in the presence Cl⁺ and Cl. The y-axis represents the etching rate measured with increasing $SiCl_2$ fluxes. Etching of the polysilicon by Cl⁺ or Cl⁺ and Cl was suppressed significantly by $SiCl_2$; however, the apparent deposition probability of $SiCl_2$ with Cl⁺ was on the order of 0.01, and with Cl⁺ + Cl was 0.05. This indicates that the apparent sticking probability of $SiCl_2$ was greatly reduced by the chlorination and/or other modification of the surface caused by the Cl⁺ and Cl fluxes.

A phenomenological model that accounts for the energy, flux and angular dependencies of Cl⁺ ion-enhanced polysilicon etching with Cl was constructed for

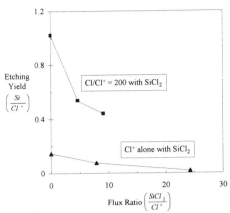

Fig. 25. The effect of $SiCl_2$ on chlorine ion-enhanced etching yield in the presence of Cl^+ and in the presence Cl^+ and Cl.

Table 2. Simplified phenomenological surface kinetics model for Cl^+ ion-enhanced polysilicon etching of polysilicon with Cl.

$$Cl_{(g)} + * \xrightarrow{\ s\ } Cl_{(s)}$$

$$Cl^+_{(g)} + * \xrightarrow{\ c(\phi)\ } Cl_{(s)}$$

$$Si_{(s)} + 4Cl_{(s)} \xrightarrow{\ \beta, c(\phi)Cl^+\ } SiCl_{4(g)} + 4*$$

use in profile simulators. Since many reaction mechanisms occur simultaneously and are convoluted, a simplified model was used to represent the overall kinetics and is presented in Table 2.

The overall sticking coefficient of Cl is s, β is the ion-enhanced reaction probability, Y_0 is the sputtering yield of Cl^+, the ion flux. The energy, flux and angular dependencies of Cl^+ ion-enhanced polysilicon etching with chlorine can be represented by the following equation:

$$\begin{aligned} Y_{total} &= C(\phi) \cdot \left[Y_0 + Y_i \right] \\ &= C(\phi) \cdot \left[Y_0 \cdot (1 - \theta_{Cl}) + \beta \cdot \theta_{Cl} \right] \end{aligned} \qquad (6)$$

where $C(\phi)$ is a constant representing the angular dependence, ϕ is the ion incident angle, and θ, the surface coverage of chlorine, is a function of the neutral to ion flux ratio. All three parameters, s, β, and Y_o, of the model scale linearly with the square root of ion energy and can be incorporated into a profile simulator to predict feature evolution in high-density plasma etching processes. A detailed description of the model can be found in JVST A 16(1), 217 (1998) by Chang et. al..

III. LOADING

It is generally observed that in a plasma etching process the etching rate drops as the etching area exposed to the plasma increases. This decrease in etching rate is true for processes that consume a major portion of the reactive species created in the plasma. Processes that are limited by the surface reaction kinetics do not exhibit this behavior. In processes where a loading effect is observed, the etching rate is typically proportional to the concentration of the reactant. For the case of etching m wafers with a plasma that produces single etching species, the loading effect is given as

$$\frac{R_o}{R_m} = 1 + m\phi = 1 + \frac{A_w k_w}{A k_s} \qquad (7)$$

where R_o is the etch rate in an empty reactor, R_m is the etch rate with m wafers present. Note that A and A_w are areas of the reactor surface and a wafer respectively, while k_w and k_s are the rate coefficients for reactant recombination at the reactors surfaces and of the reaction at the wafer respectively. Therefore ϕ is the slope of the loading curve as shown schematically in Fig. 26.

Flamm et. al. have shown that a two etchant model

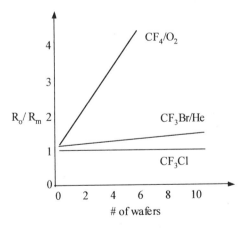

Fig. 26. Loading effect for polysilicon etching using CF_4/O_2, CF_3Br/He, and CF_3Cl plasmas (Flamm).

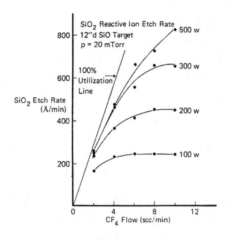

Fig. 27. SiO₂ RIE etching rate versus CF₄ flow rate.

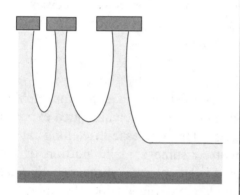

Fig. 28. Micro-loading effect.

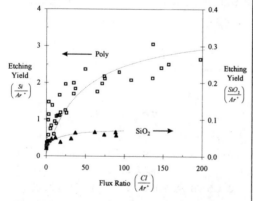

Fig. 29. Etching yield of polysilicon and silicon dioxide by Ar^+ and Cl as a function of Cl to Ar^+ flux ratio.

is necessary when dealing with etchants such as ClF_3, in which there are appreciable concentrations of both F and Cl. As can be seen from Fig. 27, the loading effect can be minimized by increasing the amount of recombinant surface or by using a better recombinant surface, but with a sacrifice in etching rate.

Increasing the gas flow rate can reduce the loading effect if the gas utilization is high, i.e., if the amount of material that can be etched, given the stoichiometry of the reaction and the flow rate, is approaching that given by the product of the etching rate and the exposed area. Shown in Fig. 27 is a set of curves that demonstrate the effect of flow rate on etching. It should be noted that if the utilization is high, it may show up as a lack of dependence on power. At very low flows, a typical process will respond linearly with flow rate at a given power, with higher flows, the etching rate of increase will become slower and go through a maximum. At very high flows, etching rate decreases, as un-reacted species are carried away.

Similarly, micro-loading is also observed in the microelectronics fabrication, which refers to the different etching rate across a wafer when there are densely populated lines and isolated lines coexisting on the wafer, exposing to the plasma. Similar to the well known effect that the average etch rate depends on how many silicon wafers that have to be etched, the area of exposed silicon at a local scale causes the variation in the etching rate, as shown in Fig. 28. The differences in local pattern density will make one area of a wafer etch at a different rate than others, and it is necessary to take these effects into consideration when designing the device layouts.

IV. SELECTIVITY

Plasma is seldom used to etch blanket films in microelectronics fabrication. The etching of photoresist and the underlying substrate is also important since the etching selectivity is never infinite and aspect ratio dependent etching often causes some etching of the underlying substrate or thin film.

Using the etching of polysilicon gate as an example, the etching will result in some etching of photoresist and needs to stop on the thin gate oxide (SiO_2). The etching selectivity between Si and SiO_2 measured by 100 eV Ar^+ and Cl is approximately 30, as shown in Fig. 29. Higher selectivities for polysilicon with respect to silicon dioxide or photoresist are desired for patterning finer features. Addition of 1% oxygen to a chlorine plasma is found to

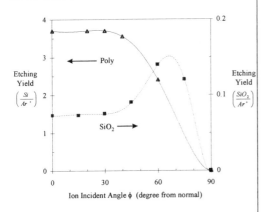

Fig. 30. Etching yield of polysilicon and silicon dioxide by 100 eV Ar^+ and Cl as a function of ion incident angles.

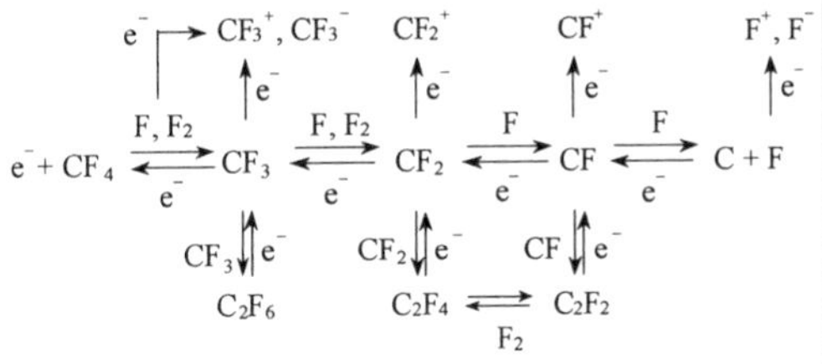

$$e^- + CF_4 \rightleftharpoons CF_3 \rightleftharpoons CF_2 \rightleftharpoons CF \rightleftharpoons C + F$$

Fig. 31. Abbreviated reaction scheme for CF_4 discharge (Kushner).

reduce the etching rate of silicon dioxide significantly, and increase the selectivity of polysilicon over silicon dioxide to 70.

The angular dependence of etching polysilicon differs from that of SiO_2 with 100 eV Ar^+ and Cl, as shown in Fig. 30. The etching yield measurements are taken at a constant flux ratio where the etching yield of either Si or SiO_2 is considered "saturated". The etching yield of polysilicon at a flux ratio of 600 exhibits no significant dependence on the angle of incidence from normal to 40° off-normal, but decreases by 35% at the angle of 60° off-normal. Maximum etching yield at near normal ion incident angles is attributed to the rapid implantation of reactive atoms into the substrate with normally incident ions. The normally incident ions consequently create mixing of the absorbed surface atoms into the lattice, induce surface chlorination, and achieve maximum etching yield at near normal ion incident angles. However, etching of silicon dioxide is mainly ion-driven, as chlorine incorporation into the SiO_2 film is limited. Ions physically sputter oxygen and silicon, allow chlorine to react to silicon, and achieve subsequent slight enhancement in the etching yield. It is necessary to incorporate proper angular dependence for each material into a profile simulator to model the surface topography evolution during the plasma etching processes.

V. DETAILED REACTION MODELING

To understand the complex reactions in the plasma, Kushner constructed "complete" kinetic models for the etching of Si and SiO_2 in C_nF_m/H_2 and C_nF_m/O_2 plasmas, by balancing the surface reactions. He was able to fit the experimental observations for these systems; however, the model included a large number of reactions with unknown rate coefficients that were estimated. Shown in Fig. 31 is an abbreviated reaction scheme for a CF_4 discharge.

The kinetic modeling approach is a useful research technique, but it suffers from a lack of appropriate rate coefficients. To be successful, the necessary rate coefficients must be determined and the surface kinetics must be developed. Probe measurements, such as optical emission or mass spectrometry, have the capability to measure the concentration of the gaseous species as a function of operating conditions, and therefore could be used to experimentally determine effective rate coefficients for many of the important reactions. The rate coefficients would be determined by fitting reaction models that adequately describe the data. Although such

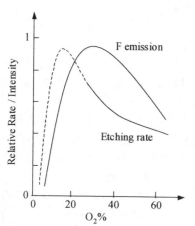

Fig. 32. The etching rate of Si and 703.7 nm emission from excited F as a function of the oxygen concentration in a CF$_4$/O$_2$ reactor.

a model does not necessarily give a mechanistically accurate picture of the plasma chemistry, it can be used to make some predictive calculations and to optimize the process efficiently. This approach has been used thus far to explain qualitatively restricted phenomena that have been observed.

For example, Flamm has used optical emission to identify F as the primary etchant species for Si in a CF$_4$ plasma and to explain the rate variation with the introduction of O$_2$. He explained the increased etching with O$_2$ addition by its reaction with CF$_n$ radicals in the plasma to produce F for etching. The optical emission used to support this claim shows a correspondence of the etching rate with the F emission (schematically shown in Fig 32). The reduction of emission with high O$_2$ concentrations was explained in terms of dilution of the plasma.

Flamm developed a simplistic, mechanistic framework to correlate the free radical chemistry of halocarbon/oxidant plasmas. Mechanistically, it offers an easy way to remember the dominant atomic species which exists in plasma with two or more halogens. This model is based upon the concept that three types of species exist in this etching system:

- Saturates: species such as CF$_3$
- Un-saturates: species such as CF$_2$
- Atoms/oxidants: the etchants species which react with the saturates, un-saturates, and surface

In the presence of unsaturates, deposition occurs. The model is formulated as

1. e$^-$ + halocarbon $\rightarrow$ saturates + unsaturates + atoms

   (e.g., e$^-$ + CF$_3$Br $\rightarrow$ CF$_3$ + CF$_2$ + Br + F)

2. atoms/molecules + unsaturates $\rightarrow$ saturates

   (e.g., F + CF$_2$ $\rightarrow$ CF$_3$)

3. atoms + surface $\rightarrow$ chemisorption or volatile products
   (e.g., F + Si* $\rightarrow$ Fsi* $\rightarrow$ SiF$_4$)

4. unsaturates + surfaces + initiating radicals $\rightarrow$ films
   (e.g., CF$_2$ + Si $\rightarrow$ SiCF$_2$* + CF$_2$ $\rightarrow$ SiCF$_2$CF$_2$ ......)

For example, in the case of a CF$_3$Br plasma, both F and Br atoms are formed by electron impact dissociation. The F atoms produced would be depleted from the plasma more rapidly than the Br through reaction with CF$_2$ leaving the predominant atomic species to be Br. The addition of oxygen to a CF$_4$ plasma reduces the number of unsaturates

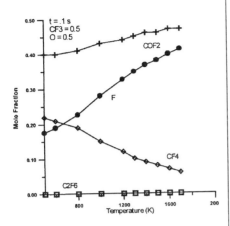

Fig. 33. Computation of products of fluorocarbon + oxygen discharge.

and inhibits the formation of a polymer film. The oxygen will also replace some of the fluorine that is bonded to carbon-bearing species, and the concentration of F increases. If large enough amounts of oxygen are added, the unsaturates can be consumed to the point that O is present as well as F within the plasma. At these levels (> 25%), the O can adsorb on a surface, forming an oxide.

This model correctly predicts the atomic species that is most apparent in the optical emission spectra. It also predicts the formation of polymer at lower concentrations of oxidants. However, it has been shown that the reactions do not occur in the gas phase, rather, the surface processes dominate. In addition, it has been shown that CF_2 is often present in high concentrations because of its low reaction probability. CF_2 is technically not a free radical, the carbon rehybridizes to fill three orbitals and while one is completely vacant of electrons. To react, the carbon must rehybridize, making its sticking probability for formation of polymer much slower than CF or CF_3. Many papers in the literature have hypothesized that CF_2 is the dominant species for oxide etching because it has such a large concentration. In fact, it has a high concentration because it reacts very slowly! The main polymer forming species are typically CF_3 and CF.

In the past few years, reaction sets for CF_4 and other chemistries have been developed. Although they are not necessarily complete, they typically are able to identify and explain trends observed experimentally. Databases by NIST and others have been compiled and kinetic codes such as ChemKin III are available in the public domain. Examples of these results for $C_2F_6 + O_2$ discharges are shown in Fig. 33. This result explains the observation that CF_4 is formed in fluorine-carbon containing discharges for typical process conditions in which the neutral gas temperature is low. At higher neutral gas temperatures, the formation of CF_4 can be avoided. This result is caused by a combination of thermodynamics which favors the formation of CF_4 at lower temperatures and COF_2 at high temperatures, as well as the kinetics of gas and surface phase reactions in which CF_x, O, and F radicals recombine.

PRINCIPLES OF PLASMA PROCESSING
Course Notes: Prof. J. P. Chang

PART B7: FEATURE EVOLUTION AND MODELING

I. FUNDAMENTALS OF FEATURE EVOLUTION IN PLASMA ETCHING

Plasma processes etch anisotropically and are used to transfer patterns from photoresist to the underlying thin film materials or substrates. There are many important parameters in plasma etching, as listed below. The greatest challenge in patterning features for microelectronics application is that each parameter can usually only be optimized at the expense of at least one of the others.

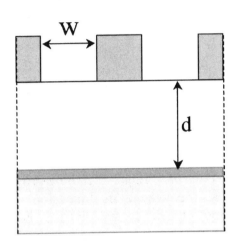

Fig. 1. Patterned thin film.

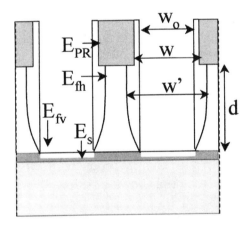

Fig. 2. Pattern transfer and bias.

1. *Critical dimension (CD) uniformity.* Uniformity across the wafer -- including densely populated areas and large open spaces, and within high aspect ratio features -- is critical to maintain consistent device performance. The uniformity is generally defined as the maximum difference in etching rates over a wafer and/or within an etcher or from batch to batch. Dividing by the average etching rate most often normalizes the variation. Processes that are categorized as "good" have uniformities in all of the above categories of less than ± 3%.

2. *Anisotropy.* It's usually desirable to have an anisotropic profile, where the etched feature edges are close to vertical, to maximize packing density on the chip. The etching anisotropy is defined as the ratio of the vertical dimension change to the horizontal change:

$$A = \frac{d}{0.5(w' - w_o)}$$

as shown in Fig. 1 and 2. A perfect isotropic etching would have an anisotropy of 1, while processes with no horizontal etching have an infinite anisotropy.

3. *Selectivity.* Defined as the ratio of the etch rate of one material versus that of another. The selectivity of thin film (f) to photoresist (PR) and the underlying substrate (s) or another film (Fig. 2).

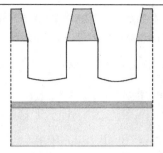

Fig. 3. Etching non-ideality: faceting.

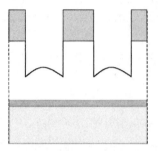

Fig. 4. Etching non-ideality: trenching.

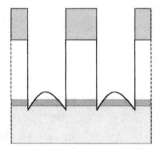

Fig. 5. Etching non-ideality: micro-trenching through the underlying layer.

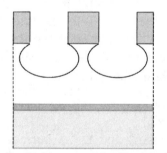

Fig. 6. Etching non-ideality: bowing.

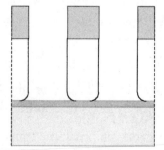

Fig. 7. Etching non-ideality: notching.

$$s_{f\_PR} = \frac{E_f}{E_{PR}}$$

$$s_{f\_s} = \frac{E_f}{E_s}$$

Note that the etching rate of the thin film has two components: the horizontal and vertical etching rates. The selectivity of the material being etched to the overlying masking photoresist layer is usually of the most concern, since this impacts CD and profile control, and the thickness of resist required (thinner photoresist is desired to adequately resolve smaller feature sizes, so selectivity must increase as geometry shrinks).

4. *High etch rate*. It is needed to keep the throughput of the system or process module high (usually measured in Angstroms/min). There is usually a tradeoff between etch rate and other parameters, such as selectivity and damage.

5. *Etch profile control*. "Bias" in etching processes is defined as the change in dimension of the feature being etched caused by the lack of infinite anisotropy, over-etching, and/or resist etching. Other etching anomalies include faceting, trenching, microtrenching, bowing, and notching, as shown in Fig. 3-7.

6. *Damage*. Plasma damage is an obvious concern, especially during gate stack formation, which will be detailed later in the Epilogue.

7. *Sidewall passivation*. Passivation is important both during and after the etch. Carbon from the photoresist mask typically combines with etching gases and etch byproducts to form a polymer-like material on the sidewall of the feature (Fig. 8 and 9).

8. *Residues*. Residue which coats the interior of the etch chamber is a difficult problem to avoid. In addition to requiring more frequent cleaning, residue is also a source of contamination. The most significant factors in controlling residue are temperature, bottom rf power, backside cooling and process pressure.

9. *Unwanted features*. Stringers, fences, veils, and

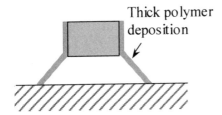

Fig. 8. Sidewall polymer deposition.

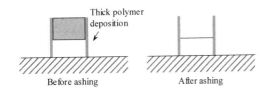

Fig. 9. Sidewall polymer depostion before and after ashing.

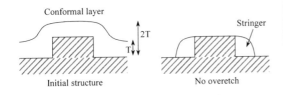

Fig. 10. Stringer formation during etching of highly conformal films.

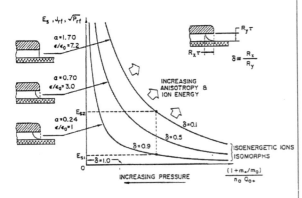

Fig. 11. Sheath potential as a function of pressure.

crowns are unwanted "decorations" that are sometimes left after an etch. For example, in patterning metal interconnect metal lines on non-planar surfaces, metal stringer may exist and short adjacent lines (Fig. 10).

10. *Corrosion.* It is mainly a problem in metal etch. Upon exposure to water vapor (i.e., air), chlorine in the photoresist will immediately attack the aluminum.

11. *Particle control.* Right now particle concentration fewer than 0.02 particles/cm^2 that are > 0.18 µm in size is required.

It has long been known that the topography of etched surface features is a function of primarily the power and pressure. For a sheath potential E_s, and a pressure p, the etching directionality is affected by mainly the ratio of E_s to n_o, the neutral density. Other parameters such as ion temperature and ion-neutral collision cross-sections are also important. It should be noted that E_s/n_o is equivalent to E_e/p in terms of the physical system. The trends of anisotropy is a function of power and pressure, as shown schematically in Fig. 11.

II. PREDICTIVE MODELING

The ever-shrinking device dimensions with corresponding higher aspect ratios have made profile control in plasma etching processes a much more difficult task. These phenomena include variation of the etching rates in reactive ion etching (RIE lag), variation of etching profile shapes (bowing, faceting, trenching), variation in selectivity to the underlying film, and variation in film morphology. At the same time, etch rates need to be maximized while minimizing the device damage to make the etching processes economically viable. Therefore, the simulation and prediction of etching profile evolution becomes increasingly important to ensure the success of a deep-submicron etching process.

Predictive profile simulation has been long sought as a means to reduce the time and cost associated with trial-and-error process development and/or equipment design. Profile control is one of the most important aspects in pattern transfer as it determines the success of subsequent deposition processes and ultimately the device performance. To date, simulation work has given invaluable insight into the surface profile evolution during ion-enhanced plasma etching, using various techniques such as string algorithm, characteristics method, shock-tracing method, and direct simulation Monte Carlo (DSMC) method. Dominant reaction mechanisms incorporated in these simulators include ion induced etching and ion reflection. More recently, localized surface charging effects and redeposition on the sidewalls have also been taken into account.

Commonly observed high-density plasma etching peculiarities such as bowing, tapering, undercutting, trenching and micro-trenching have been predicted as different physical or chemical mechanisms are incorporated. However, comparing simulation results to etching profiles can not unambiguously determine the dominant surface phenomena as the profile evolution is often a function of the difference between several mechanisms, e.g. competition between etching and deposition processes. The profiles can often be fitted using more than one set of surface kinetics. Therefore, a comprehensive understanding of the dominant plasma-surface reaction pathways is needed to develop a general, predictive profile simulator. The use of beam studies in which the fluxes are well characterized to measure unambiguously the etching kinetics is valuable in determining the functionality of reactive species at specific, well-defined etching conditions, which can be generalized to the much more complex plasma environment.

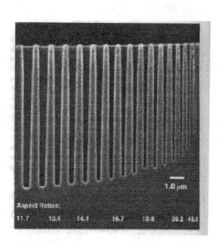

Fig. 12. Illustration of aspect ratio dependent etching.

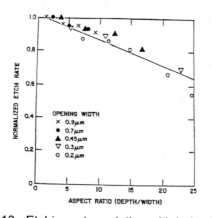

Fig. 13. Etching rate variation with hole sizes.

III. MECHANISMS OF PROFILE EVOLUTION

The etching of features is dominated by processes which are dependent upon the feature aspect ratio, i.e. the feature depth to width, as shown in Fig. 12. The etching rate of smaller features is slower compared to that of larger features. This is called Aspect Ratio Dependent Etching (ARDE) or Reactive Ion Etching lag (RIE lag), as shown in Fig. 13.

RIE lag in which features of high aspect ratio etch more slowly is typically caused by the reduction of ions or neutrals to the bottom of the feature. This reduction is a combination of both the reduced view

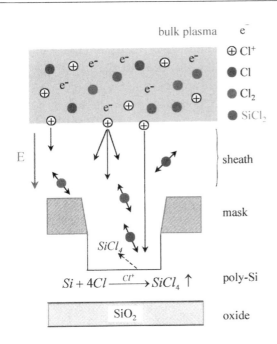

Fig. 14. Non-ideality in plasma etching: (a) ion distribution is not uni-directional and (b) photoresist is etched and re-deposited.

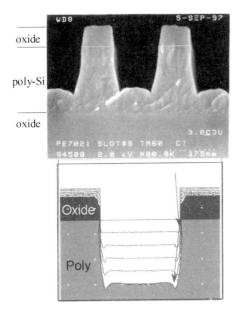

Fig. 15. Ion trajectory changes causes microtrenches.

factor of the higher aspect ratio feature and the charging within the feature. Lowering the process pressure makes the sheath more collisionless, thereby increasing the ion directionality. The greater ion directionality reduces the dependence of the flux on the view factor. Increasing the ion energy also increases the ion directionality in a high-density, low-pressure plasma. The ion energy dispersion is determined by the collisions in the presheath. Greater acceleration in the sheath results in an increased directionality. The effect of surface charging on the ion flux to the bottom is determined by the aspect ratio of the feature and the ion to electron energy ratio. The greater the energy ratio, the less charging there is.

Since all of the above mentioned phenomena scales with the feature aspect ratio, the etching profiles are typically modeled to reduce the number of experiments that must be performed and to project the effects to other sized features. As the minimum line width decreases RIE Lag and Inverse RIE Lag, in which small features etch more rapidly, become more significant. The reduction of RIE lag has driven the industry to low pressure, high-density plasma sources.

Here, the important parameters affecting the evolution of features during etching and deposition are summarized:

1. Ion bombardment directionality

The ion bombardment directionality is primarily a function of the collisionality of the sheath (i.e., its thickness in terms of mean-free-path lengths) as discussed in the section on ion bombardment energy and angular dispersion. The view factor to the plasma and the ion angular/energy distribution determines the flux on each element of the feature surface. Reactive ion etch lag (RIE lag) in which smaller features (i.e. a high depth to width ratio) etch more slowly is the primarily caused by this effect.

2. Ion scattering within the feature

The energetic ions can scatter from the surface features and produce significant fluxes at other parts of the surface profile. Trenching, the more rapid etching of the surface at the base of the sidewall, is one artifact of ion scattering. If the resist does not have a vertical profile or becomes faceted, ion scattering onto the opposite sidewall and undercut-

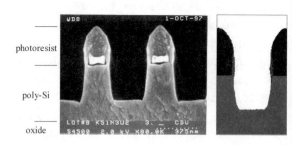

Fig. 16. Surface deposition/re-deposition modifies the etched profile.

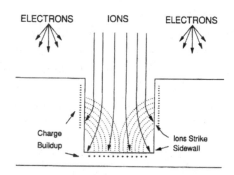

Fig. 17. Surface charging.

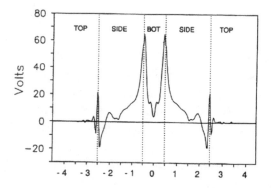

Fig. 18. Computed potential build up on the surface as a function of position for an isotropic electron flux with a temperature of 5 eV and an anisotropic ion flux of 100 eV. The potentials are for a feature with an aspect ratio of 2. Note that in this figure −0.5 and 0.5 is the bottom, ±(2.5 to 0.5) is the sidewall, and ±(4.5 to 2.5) is the top.

ting can also be observed due to ion scattering. The significance of ion scattering is both a function of the slope of the feature sidewalls during etching and the angular dispersion of the ion bombardment. Obviously, with highly directional ion bombardment and vertical features, artifacts such as trenching occur. However, with nonvertical sidewall features, ion bombardment with a greater angular dispersion can produce less trenching, because the scattered ions are more dispersed. As shown in Fig. 15, because ion scattering within the features occurs at impingement angles greater than 40° off normal of the profile surface, this scattering can lead to the formation of microtrenches near the base of a sidewall where the ion flux is the largest. The ion flux striking the feature bottom is the sum of the direct flux from the plasma and the flux reflected off the sidewalls. Simulation of the over-etch step using this non-uniform ion flux at the feature bottom, demonstrates that the oxide layer is broken through at the positions with the highest ion flux. Microtrenches are formed in the underlying silicon substrate once the thin oxide is etched through.

3. Deposition rate of passivants from the plasma

Passivation is essential in highly directional etching processes to prevent spontaneous etching as discussed earlier. The passivation can be produced by production of depositing species in the plasma such as CF which plates out on the surface features to build up a polymeric passivant layer. These can originate from the introduction of gases such as CHF_3 or the etching of photoresist. Stable volatile products such as $SiCl_4$ can also be broken up by the plasma to produce depositing species such as $SiCl_2$.

The effect of photoresist erosion was directly related to redeposition and etching directionality. The presence of excess amounts of passivant deposition leads to the narrowing of features, i.e. "tapered" in which the sidewall tapers out into the unmasked region. The thick sidewall passivation acts as a mask for the next increment of etching. The presence of such a tapered feature also affects ion scattering, which can alter the etched profiles. As shown in Fig. 16, micro-tenching is eliminated when photoresist is used as the mask (compared to a hard mask), due to the redeposition modifying the slope of sidewalls and altering the etched profile.

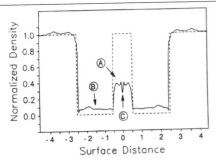

Fig. 19. Computed flux to the feature for the feature with an aspect ratio of 2. The dotted line indicates the expected flux if the charging had not occurred. Charging is shown to reduce the ion flux to the bottom of the feature and enhance the flux to the sidewalls.

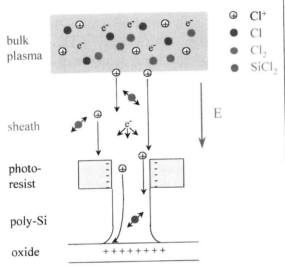

Fig. 20. Formation of notch at the end of the etching of a conducting materials on an insulating material.

4. Line-of-sight redeposition of products

The non-volatile etching products (from etching of photoresist or thin films) can also re-deposit and modify the feature profile. Line-of-sight processes can be very important in narrow features. Note that the mean free path is long with respect to any feature size in most of the plasma etching processes. Therefore processes on one surface can couple directly with those on any other surface which has a non-zero view factor. For example, the lack of sidewall passivation from sputtered photoresist from the next line can lead to undercutting and "mouse bites" on the outside sidewall of a line. A mouse bite is the breakthrough of the sidewall passivation allowing rapid spontaneous etching that produces a cavity. When observed in an SEM, it looks like a mouse bite into a line.

Excess deposition of product on the sidewall of the photoresist can lead to the formation of involatile products, e.g. SiO_2 after ashing. The removal of the photoresist, but not the sidewall passivants causes the formation of "ears" on the line, as shown previously in Fig. 9.

5. Charging of surfaces in the features

Features can become charged by two different mechanisms, as detailed in Figs. 17-19. First, in the presence of a magnetic field a plasma can support a potential gradient across a wafer. As the wafer is conductive or the film being etched may be conductive, the potential of the conductive wafer will be uniform. The photoresist and/or other insulating surfaces can charge to different potentials where the net flux of ions and electrons from the plasma is neutral, if little surface leakage or bulk film conductivity occurs. Under these conditions, the charging of the features can lead to different etching profile across the wafer.

A second mechanism for charging is caused by the differing angular dispersion of the ions and the electrons. The positive ions are highly directional compared to the electrons which act as a free gas and would be "isotropic". Since the electrons are captured with a probability approaching unity, their true distribution is near that of a gas incident on a surface which is cosine with respect to the surface normal. The steady state condition must have a net equal flux of electrons and ions for a perfectly

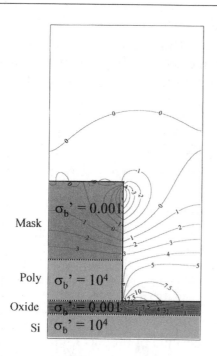

Fig. 21. Potential profile in a feature with an "insulating" bottom and "conducting" poly sidewalls. Potentials approaching ion energy develop at feature bottom corner; negative potentials at top corner.

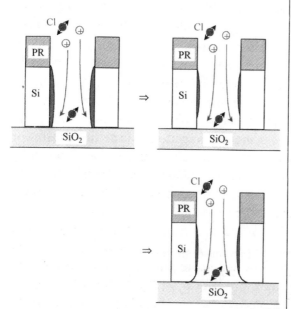

Fig. 22. The passivation material barrier is removed and spontaneous chemical etching occurs, leading to the notch formation.

insulating surface. The surface potential builds up until that is achieved. This phenomenon is a function of the geometry of the surface (aspect ratio), the ion directionality, the ion energy, and the electron energy. Because of the differing flux at the edge compared to the interior of a set of lines and spaces, the edge lines can etch differently. Often the edge line will receive more passivant on the outside of the line because of the larger view factor for neutral passivant deposition. This mitigates the effect of the excess ion flux on the outside of the line, but can lead to excess flux on the inside sidewall of the outside line.

In addition, it is particularly important for conditions with lower ion bombardment energy and higher electron energy which is characteristic of most new high density plasma sources. The charging can be several times that of the electron energy for aspect ratios of 2 and therefore can be a few tens of volts. As the potential can build up to a few tens of volts, this charging may also contribute to gate oxide rupture. In gate oxide rupture, the potential across a gate during an etching process exceeds a threshold value and current passes, thereby damaging the oxide.

Surface charging can also contribute to RIE lag, trenching, and barrel etching artifacts. In addition, the increased flux of ions at the foot of the sidewall can lead to rapid undercutting. Thinning of the passivant at the bottom because of charging also may lead to the spontaneous etching of the silicon-oxide interface, causing the formation of a notch. The chemical etching is more rapid because the stress at the interface accelerates the rate of spontaneous etching, and the high stress state of the interface is believed to cause spontaneous etching in undoped and p-type polysilicon (Fig. 20-22).

IV. PROFILE SIMULATION

Numerous models for profile evolution in plasma etching and deposition can be found in the literature. All of these models depend upon the user entering into the model, information for the etching or deposition kinetics, e.g. ion enhanced etching rate, ion directionality, spontaneous deposition rate, sticking probability of reactants, etc. Various algorithms are then used to compute the movement of the surface with time and display it. The most

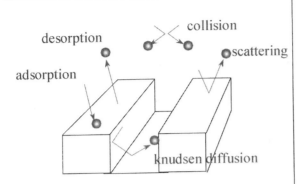

Fig. 23. Concept of DSMC.

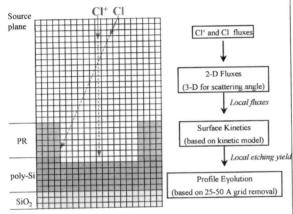

Fig. 24. Schematic drawing of Monte Carlo simulation

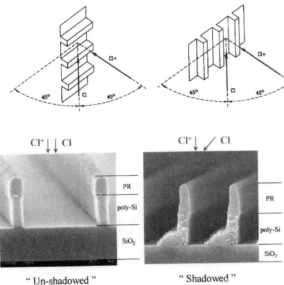

"Un-shadowed"   "Shadowed"

Fig. 25. Etched profiles for simulation evaluation.

common programs in use are 3-D Simulator (UIUC), SAMPLE (Berkeley), EVOLVE (RPI) and SPEEDIE (Stanford). Each of these models presently have the capabilities to include certain physical mechanisms that were discussed above.

Monte Carlo Profile Simulator

An example of Monte Carlo simulation is given here to explain the feature evolution modeling. The simulation domain spanned the centerlines of neighboring photoresist lines that defined the trench to be etched (Fig. 23 and 24). The domain was discretized into square cells with 50Å sides, a dimension comparable to the surface chlorination layer depth caused during ion induced etching processes. Trajectories of individual highly directional ions and isotropic neutrals were tracked from the source plane until they encountered a surface. At the surface, the reactant species particle could either scatter or stick. Using this Monte Carlo algorithm permitted us to incorporate the dominant physics and chemistry of the etching process easily by using kinetically based probabilities for each surface process and performing an elemental balance for cell each time the surface interacted with a reactant species. The surface probabilities incorporated in the simulator were based on the surface kinetics models developed. They included angle-dependent ion-enhanced etching, sputtering, ion reflection, recombination of neutrals on surfaces, deposition from the source plane and redeposition of etching by-products. Whenever a surface cell was etched or deposited, the surface was redefined.

Patterned polysilicon samples were etched in the beam apparatus to generate test structures for modeling confirmation. The samples were etched by 35 eV Cl^+ at a neutral-to-ion flux ratio of 200. The patterned samples were aligned so that the patterned lines were either parallel or perpendicular to the beam plane, as shown in Fig. 25. When the patterned lines were parallel to the beam plane, the photoresist did not shadow the beam fluxes producing uniform incident fluxes of both Cl^+ and Cl at the bottom of the feature. However, when the patterned lines were perpendicular to the beam plane, the photoresist "blocked" the Cl beam flux from reaching part or all of the feature bottom.

In the unshadowed orientation, an anisotropic etching profile was observed after etching through

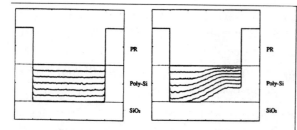

Fig. 26. The orientation of the patterned samples with respect to the beam plane, the SEM images of the etched sample in Fig. 25, and the Monte Carlo simulated profiles of (left) an un-shadowed case, and (right) a shadowed case. Profiles etched by 35 eV Cl^+ and Cl at a neutral-to-ion flux ratio of 200.

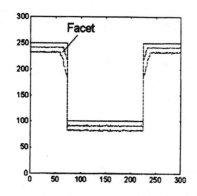

Fig. 27. Simulation of faceting caused by physical sputtering of feature.

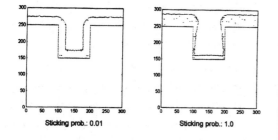

Fig. 28. Deposition in a trench from an isotropic flux with 0.01 and 1.0 sticking probabilities.

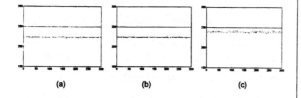

Fig. 29. Roughening of surfaces during etching

the polysilicon film to the underlying oxide film. Slight undercutting of the sidewall indicated no sidewall passivation, which is consistent with the good photoresist integrity and lack of by-product deposition in the beam apparatus.

In the shadowed orientation, a directional etching was observed at the left side of the trench where sufficient Cl^+ and Cl fluxes induced anisotropic etching of the polysilicon. The reduced Cl flux on the right side of the trench that was shadowed caused a slower etching rate resulting in unetched polysilicon. Since the etching of the polysilicon on the right side is too high to be just Cl^+ sputtering, and too low if all Cl atoms scattered from the photoresist contributed to the etching of the right side of the feature. It is postulated that chlorine atoms recombine each other to form molecular chlorine upon colliding with the photoresist surface. Since the reactivity of molecular chlorine is much lower than that of atomic chlorine on a highly chlorinated surface, we expect reduced contribution of the scattered neutrals to the etching of the polysilicon on the right corner. By carrying out a sensitivity analysis for the recombination probability of Cl on the photoresist, it is found that a probability of 0.5 best represented the measured profile.

To test the Monte Carlo simulator, several test cases were performed in which a subset of all possible surface processes were allowed to occur, the faceting of resist caused by physical sputtering is shown in Fig. 27. The photoresist lines were subjected to directional ion bombardment with no redeposition of the sputtered species. The highly directional ion flux produces "faceting" in which the facet angle is equal to the angle corresponding to the maximum sputtering yield. The Monte Carlo simulator captures this peculiar effect with the facet being formed at 63° from the horizontal.

Plasma etching processes typically include deposition processes from either the plasma or redeposition of etched products. This deposition passivates the sidewalls preventing undercutting and modifies the etching profile evolution. The Monte Carlo simulator can predict deposition profiles under a variety of conditions. Shown in Fig. 28 are two sample profiles with an isotropic flux of depositing species from the gas phase on a trench with a unity aspect ratio. For low sticking probabilities, the deposition is conformal. However, with unity sticking probability, highly non-conformal deposition is observed.

Figure 29 also demonstrates that the roughening of surfaces during plasma etching can be simulated; a) shows the profile obtained by sputtering a planar surface in which the sputter products are not allowed to redeposit. The surface roughening is caused by the statistical variation of the ion flux and the variation in sputtering yield with angle (the same effect which causes faceting). Part b) was obtained under the same conditions but allowing the products to redeposit. Though very similar to a), one can see more order in the small pillar-like structures on the surface and a greater roughness. Redeposition lowers the etching rate of the surface slightly. In part c) deposition from the source plane is included and grass-like structures are observed. The tips of the surface features receive a greater deposition flux because of their larger view factor of the plasma while the valleys are shielded and therefore have a lower deposition flux. Another mechanism for grass formation involves micromasking of the surface that this simulator can also model.

V. PLASMA DAMAGE

The exposure of wafers to a plasma causes a variety of "damage" to the thin films beneath, as summarized in Table 1 and 2 at the end of this section. The nature of these damage mechanisms is varied but are all considered faulty during electrical testing.

1. Contamination

First, metals sputtered from the electrodes or walls of the chamber can deposit on the wafer, leading to contamination. This source of damage can be dealt with by reducing the plasma potential with respect to the chamber surface until it is less than the sputtering threshold, typically about 15 eV. In high-density etchers, the self-bias of the plasma potential is approximately 25 eV, thus the electrode on which the wafer rests must also be regarded as a possible source. One solution is to make all the surfaces that can be sputtered of high purity, non-objectionable materials; typically anodized aluminum.

Contaminants from the ashing of photoresist, in particular Na and K can be problematic as they are mobile ions in oxide and can cause the threshold voltage shift of MOSFETs. Wet cleans are usually effective in removing these contaminations. Some studies have indicated that the exposure of the wafer to ion bombardment and/or UV results in the incorporation of these alkali metals into oxide making it difficult to be removed by a short wet clean.

2. Particulates

The dominant loss process in all microelectronics processing is point defects that are typically attributable to particulates contamination. The major mechanisms for particulate contamination in plasma processes are mechanical abrasion of moving robotic parts, the flaking of deposited films onto the wafer, degradation of materials, arcing, and the formation of particulates in plasma phase.

The formation of particulates by mechanical abrasion is obviously most readily avoided by avoiding abrasion and/or keeping any parts that must come in contact with the backside of the wafer. Limiting the flow rate of venting gas in the robot has also been found to reduce the levitation of particulates from the bottom of the chamber. In certain processes, the rapid pump down of chambers leads to adiabatic cooling of air and can induce water particulates that in turn deposit on the wafers.

Materials that slowly react with the plasma products can degrade and produce particulates. In particular, O-rings that are used in seals, if made of an incompatible material, can shed particulates after plasma exposure.

In plasma processes, arcs can occur at sharp corners because of the high electric fields caused by the small radius of curvature. The rounding of all edges (sometimes due to welding) reduces arcing and associated evaporation of material near the arc, and subsequent particulate formation.

In etching processes, particulates can be formed in the gas phase under extreme conditions and their motions are governed by the sheath dynamics. In PECVD particulate formation in the gas phase is always a concern. The plasma charges particulates negatively, thereby suspending them. They undergo transport by thermal gradients, ion wind, concentration gradient, gas flow, electric field, and many other forces. These particulates can be eliminated from the wafer region by certain gas flow

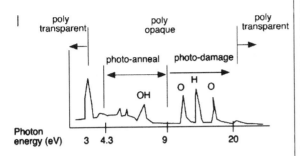

Fig. 30. The origin of photon induced damage to ultra-thin dielectrics.

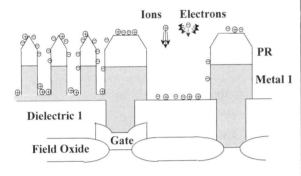

Fig. 31. Electrical stress induced damage.

patterns that overcome the dominating electric fields. The formation can be curtailed before the particles grow to a size that is problematic by pulsing the plasma. In the absence of the plasma, the small particles are sweep out of the reactor with the gas flow.

3. Gate oxide damage – photon

Transistor gate oxides can also be damaged by photons with energies greater than about 10 eV, which can excite electrons into the conduction band of silicon dioxide from the valence band, leaving behind a trapped holes (broken bond) and probably creating damage. If the oxide is a gate dielectric layer for the MOSFET, reliability may be impaired. Low-energy photons (4-9 eV) excite electrons into the oxide to neutralize holes created by higher-energy ion bombardment and can reduce the damage. Photon energies higher than 16-20 eV can penetrate polysilicon or metal layers, damaging the gate oxides even when they are covered by gate metal. The net result in terms of oxide damage involves a complex interaction between the spectrum of the plasma and the properties of the layers, as depicted schematically in Fig. 30. Normally, once a circuit has received a thick layer of CVD oxide dielectric, the gate oxide is well-protected from further radiation damage.

4. Gate oxide damage – electrical stress

The potential variations on a wafer during processing can connect to individual gates and if the voltage exceeds 10-20 volts, current passes through the gate oxide causing the gate oxide to degrade. Electrical shorting and low gate breakdown can be observed after such process induced electrical stress (Fig. 31).

It is known that such damage occurs after a film is cleared. It is not confined to gate etch, but is actually most likely in second metal etch. The metal lines electrically connect to individual gates and thus can induce gate oxide damage without direct plasma exposure. Much work has been done with test structures which have striven to ascribe the damage to one of two mechanisms: 1) an *area ratio effect* in which the large conductive line serves to collect excess charge and funnel it to a single gate or 2) a *perimeter effect* in which the charging occurs because of the perimeter of the conductive line collecting charge.

The effects appear to be caused by a combination of the isotropic electron and anisotropic ion fluxes that lead to charging within a feature. The perimeter of the lines has a larger view to the electrons than a line surrounded by other lines. The outside line, therefore, charges to a different potential than inside lines. This leads to both different feature profiles as well as potentially electrical damage.

5. Lattice damage

The ion bombardment during etching can induce lattice damage and the incorporation of hydrogen and fluorine quite deep into the silicon substrate. Ion bombardment also produces a large flux of interstitials that can diffuse into the lattice. It has been observed that pn junction below a depth of about 60 nm are not affected, but shallower junction can exhibit leakage and non-idealities. The incorporation of carbon into the first few monolayers also occurs and can increase the resistance. The reduction of ion bombardment energy in general reduces these effects.

6. Post-etch corrosion

Post-etch corrosion is problematic after aluminum metal etches as aluminum undergoes electrochemical attach. Chlorine left on the wafer surface after etching is absorbed in a condensed water layer upon exposure to air forming a thin electrolyte solution on the surface. The corrosion is particular problematic when the aluminum is in contact with a different metals such as TiW which is used as a barrier layer. The presence of a second metal forms a "battery" which discharges by the corrosion of the aluminum. As the technology evolves, since the new interconnect material, Cu, can not be etched, the challenges are on the etching of low-k materials and their integration.

Table 1. Potential damage inherent in plasma-based processing (IBM)

Damage type	Basic cause	Present in etching or deposition	Materials affected
Residue contamination	Exposure to plasma	Usually only in etching due to reaction by-products remaining on surfaces	All
Plasma-caused species permeation	Exposure to plasma	Both	Dielectrics and semiconductors
Bonding disruption	Exposure to plasma Particle and/or photon bombardment	Both	Dielectrics and semiconductors
Current flow damage	Current flow during plasma processing due to charging or induced EMFs	Both	Dielectrics

Table 2   Impact of processing flow and of device and circuit layout (IBM)

Impact category	Plasma damage type affected	Example
Processing flow: Passivation	Bonding disruption due to bombardment Wear-out damage	Hydrogen released in subsequent processing such as a postmetallization anneal or a hydrogen-laden plasma step can passivate damage
Processing flow: Activation	Bonding disruption due to bombardment Current flow damage	Charging current of a subsequent plasma-based step can activate dielectric damage passivated in an earlier step
Processing flow: Cumulative effects	Bonding disruption due to bombardment Current flow damage	Dielectric current flow damage of a plasma-based step can augment current flow damage of an earlier plasma-based step
Layout: Antenna effects	Current flow damage	Interconnects can collect charge that must pass through a dielectric to dissipate
Layout: EMF loop effects	Current flow damage	Interconnect loops can induce EMFs from time-changing magnetic fields, thereby setting up currents across dielectrics
Layout: Edge effects	Bonding disruption due to bombardment Current flow damage	Device edge exposure to plasmas can cause damage due to bombardment and/or increased current flow damage arising from edge conduction

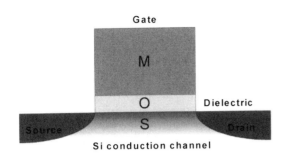

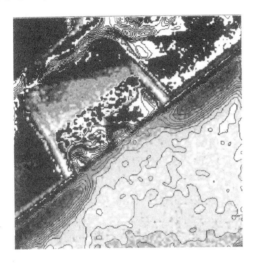

Fig. 1: A thin layer of SiO₂ or other dielectric separates the gate electrode from the conduction layer in a MOS transistor.

Fig. 2. The electrostatic potential in an 0.35 μm MOS transistor, mapped by electron holography, showing the actual structure in an 0.7 × 0.7 μm field [Gribelyuk et al., Phys. Rev. Lett. **89**, 025502 (2002)].

PRINCIPLES OF PLASMA PROCESSING

EPILOGUE

CURRENT PROBLEMS IN SEMICONDUCTOR PROCESSING

In a rapidly changing technology, problems are encountered and solved continually. Though the general principles covered in the main lectures will not change from year to year, the material in this section is just one snapshot in time and should be updated yearly. Most areas of research fall into the categories of front-end design, interconnects, oxide damage, and species control in plasma sources.

Design of smaller transistors has led to the need for high-k dielectrics and ways to etch them. Interconnects are metal conductors, surrounded by insulators, used for electrical connections within the chip. To speed up the signals along these paths, the RC time constant has to be reduced; thus the need for lower-resistance metals and low-k dielectrics. Efficient oxide etchers for the damascene process are used to address this problem. Plasma-induced damage to thin gate insulators is such a severe problem that it can control the type of plasma reactor chosen for etching. Plasma sources for processing the next generation of 300-mm wafers are already in production, and new types are not yet needed. However, measurement and control of the atomic and molecular species impinging on the wafer is still a problem in existing devices.

I. FRONT-END CHALLENGES

1. High-k dielectrics

The basic building block of a CPU (central processing unit) is the MOSFET transistor, whose simplified structure is shown in Figs. 1 and 2. If we want faster devices while maintaining reasonable drain current, the channel length has to be shorter. However, if the channel length is decreased, the gate area will also decrease, lowering its capacitive coupling to the channel. So the gate dielectric thickness has to be decreased to maintain the same gate capacitance. However, SiO₂ layers cannot be made much thinner than about 20Å because of leakage currents and the difficulty of maintaining uniform thickness. With high-k materials, the dielectric constant

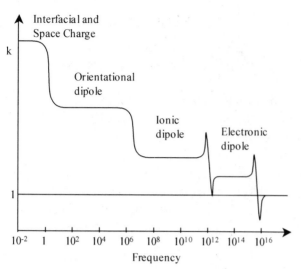

Fig. 3: Dielectric constant k vs. frequency.

Table 1. Potential high-k candidates.

Material	κ	E$_{BD}$(MV/cm)	E$_G$ (eV)
SiO$_2$	4	15	8
SiO$_x$N$_y$	4	15	6
Si$_3$N$_4$	5-7	10-11	5
TiO$_2$	100	0.5	4
Ta$_2$O$_5$	25	4	4
Al$_2$O$_3$	11	10	8
ZrO$_2$	22	15	7
HfO$_2$	22	15	7
Y$_2$O$_3$	15	5	6

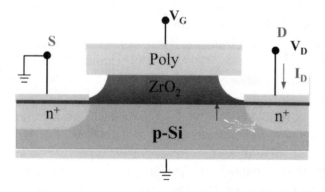

Figure 4. Etching ZrO$_2$ has problems with undercutting and incomplete removal of a silicate layer below the source and drain contacts.

can be increased to maintain the same capacitance. Power consumption will also be reduced by the lower voltage and current requirements of these down-sized devices.

At frequencies of 1–1000 GHz, the dielectric constant of the insulating material will decrease because only the ionic and electronic dipoles respond to the applied field, as shown in Figure 3. Since the ionic component of the polarization arises from the relative motion of the charged atoms and the electronic component stems from distortion of the electron cloud, high-Z dielectrics, such as most of the metal oxides, tend to have a higher k than SiO$_2$ and thus are considered as alternative gate dielectric materials. Moreover, as the physical thickness of the dielectric layer decreases below 10Å, direct quantum-mechanical tunneling causes an unacceptably high leakage current, and breakdown of the dielectric occurs at unacceptably low voltages. A thicker dielectric film with a higher k will also alleviate this problem. For example, ZrO$_2$, HfO$_2$, and their silicates are being widely researched as alternative gate dielectric materials (Table 1).

It is essential to develop an anisotropic patterning process for these high-k dielectric materials since: a) these high-k dielectric thin films are thicker and more chemically resistant to HF compared to SiO$_2$, so that isotropic etching in HF would undercut the gate dielectric material more and affect the device reliability; and b) some of the promising high-k materials such as zirconium silicate (ZrSi$_x$O$_y$) are quite inert to strong acids, making it difficult to completely remove the high-k materials above the source and drain regions of the transistor. It has been shown that less than a monolayer coverage of ZrSi$_x$O$_y$ would result in a high contact resistance and significantly reduced current. Figure 4 shows a schematic diagram demonstrating the issues of incomplete removal of interfacial layer and undercutting.

2. Metal gates

Highly doped polysilicon is used for the gate metal with SiO$_2$, but migration of the dopant causes a depleted layer which decreases the gate capacitance, requiring the oxide layer to be even thinner. High-k insulators alleviate this problem because they can be more compatible with elemental and compound gate metals. Metal gates would also help eliminate possible reactions between poly-Si and the high-k materi-

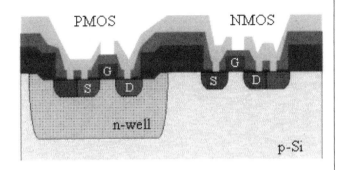

Fig. 5. Single level metallization: the top layer is the metal.

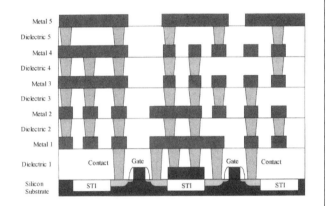

Fig. 6: Multiple level metallization. Connections to seemingly isolated blocks are made in planes in front of or behind this plane.

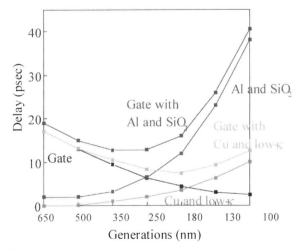

Fig. 7. Gate delay and RC time constant, and their sum, for new and old gate materials.

als.

The search for an alternative gate electrode relies largely on the work function analysis of the potential gate material, process compatibility with dielectric deposition and annealing, and thermal and chemical interface stability with dielectrics. To replace n^+ and p^+ polysilicon and maintain the scaled performance for 50 nm CMOS and beyond, metals and their nitrides, silicides, and oxides, such as Mo, Hf, MoN, HfN, Pt, Ni, and RuO_2, are all being researched as potential gate materials. Of course, proper etching chemistries have to be identified and studied to allow the patterning of these electrodes.

II. BACK-END CHALLENGES

To complete the formation of an integrated circuit, the solid-state devices need to be interconnected and finally get connections to the world outside the silicon chip. In this section, we will discuss metallization and interconnection isolation using dielectric materials, and the challenges for future interconnects.

Metals or heavily doped polysilicon have been used to wire the devices, and it is important that these "wires" have low resistance, make good ohmic contacts, and be properly insulated with dielectric materials. The product of the resistance R of the metal lines and the capacitance C of the dielectric materials gives rise to the RC delay in the integrated circuit. Depending upon the materials used, barrier layers are sometimes needed between the metal and the dielectric materials. The single level interconnection in a CMOS structure is shown in Fig. 5. Typically, five- to nine-level interconnection schemes are needed for a high density of devices (Fig. 6). As shown in Fig. 7, at device dimensions below 250 nm, the interconnection RC delay exceeds the gate delay (needed to establish the channel and the current in it) and dominates the speed of the circuit. This crossover prompts the search for lower resistance metal (Cu) and lower capacitance interlayer dielectric (ILD) materials. Therefore, it is critically important to address the challenging issues in metallization and interconnect isolation.

1. Copper metallization

To reduce the RC delay, copper is chosen for its lower resistivity compared to aluminum. Copper is deposited by electroplating because the process is cheap and robust. However, copper halides are not

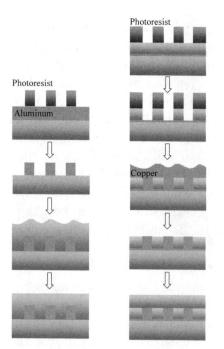

Fig. 8. Left: Conventional subtractive interconnection scheme. Right: Damascene process.

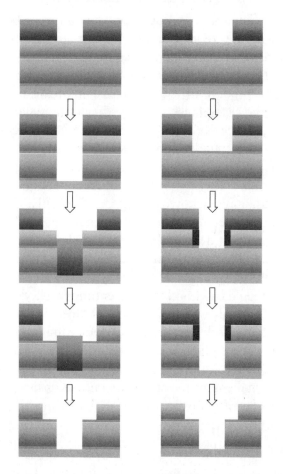

Fig. 9. (Left) via-first and (right) trench-first dual damascene processes.

very volatile, so no low temperature dry etching process is available to pattern copper. Therefore, electroplating of copper into preformed dielectrics (the *damascene* process) is emerging as the new deposition method for interconnects in advanced integrated circuits.

A conventional Al interconnect etching scheme is depicted in Fig.8 (left). First, photoresist is used to etch the metal to the desired pattern and then subsequently removed. Dielectric material is then deposited on the patterned metal, followed by chemical mechanical polishing (CMP) of the dielectric to yield a planar surface.

The word *damascene* comes from the city of Damascus, where the process of metal inlaid decoration was invented. In Fig. 8 (right), photoresist is used to etch a dielectric material to the reverse of the metal pattern desired, up to a stop layer. After ashing of the resist, a barrier layer and a Cu seed layer are deposited, and then copper is electroplated into the patterned dielectric. Chemical mechanical polishing of the metal and subsequent deposition of more dielectric produces the same structure as on the left.

From a processing perspective, electroplating is capable of void-free filling of sub-micron trenches and vias. From an interconnect reliability perspective, electroplating provides a surprising route to achieving large-grained bamboo-type interconnect structures that improve electromigration resistance.

The advantages of copper metallization are:

- Lower resistivity ($1.7\mu\Omega$–cm, vs. 2.7 for Al)
- High melting temperature (1083°C)
- Better electromigration resistance
- Inexpensive with electroplating

The disadvantages of copper metallization are:

- Barrier material needed
- No dry etching process available
- Corrosion

The challenges in patterning interlayer dielectrics for copper interconnects can be demonstrated by the two commonly used *dual damascene* processes shown in Figure 9. Metal in small vias is used at the bottom, where space is at a premium due to the high packing density of devices. The dielectric material there is low-k, because R is large. Above that layer, larger trenches can be used, possibly with SiO_2 (in

Fig. 10. Multilayer Cu interconnections, seen with the interlayer SiO₂ etched away.

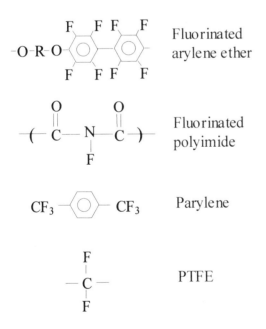

Fig. 11. Structure of organic low-k dielectric materials

red). The via-first damascene process begins with the deposition of a thick dielectric on top of a thin etch-stop layer. A deep via pattern is etched through the entire dielectric stack. Photoresist processing fills the via and generates the trench pattern. The buried etch stop is used to terminate the trench etch, and the photoresist is removed to yield the final pattern. The advantages of the via-first process include the single step dielectric etch (giving highest throughput), an easier lithography process, and the flexibility to include multi-layer dielectrics. The disadvantages are possible misalignment leading to reduced via size, selectivity variations with different dielectrics, need for high selectivity to SiN, and difficulty in removing the resist at the via bottom.

The trench-first damascene process starts the same way, but the wider trench pattern is etched in the first dielectric up to the stop layer. Another photolithography step is used to pattern and etch the via. Removal of the photoresist yields the same structure as on the left. The advantage of this process is that the two etch processes can be optimized for each dielectric material or a single thick layer of dielectric materials can be used. The disadvantages include the required timed-etch for the line and the difficulty in patterning through a thick layer of photoresist. Fig. 10 shows the complexity of the final product of multilayer copper metallization.

The buried-via (a.k.a. self-aligned dual damascene) process involves the patterning of the nitride stop layer to define the via, followed by another interlayer dielectric deposition. Thus, only one etching step is required to form the same structure shown in Figure 10 (not illustrated).

2. Interlayer dielectrics (ILDs)

Silicon dioxide (k ≥ 3.9) has traditionally been used for interconnect isolation. However, lower dielectric constant materials are needed to further reduce the RC delay. These low-k materials not only reduce the line-to-line capacitance, but also minimize cross-talk noise and reduce power consumption. A broad spectrum of low-k materials including fluorinated SiO₂, organic polymers, nanoporous silica, amorphous fluorocarbon, and hybrid inorganic and organic materials have been investigated for replacing silicon dioxide as ILD (Fig. 11). The effectiveness of new dielectric materials depend on the dielectric constant, thermal stability, water resistance,

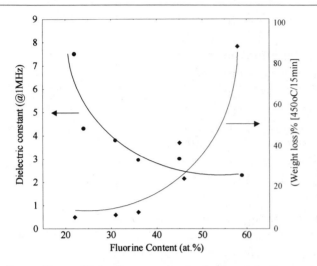

Fig. 12. Dielectic constant (left) and thermal instability (right) of a-C:F films as a function of F content. As k drops, weight loss increases.

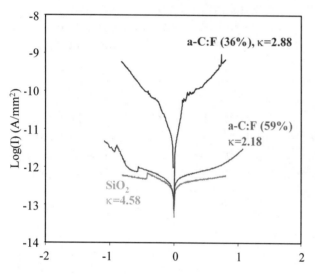

Fig. 13. Leakage current of a-C:F films vs. voltage. As k drops, the leakage current increases.

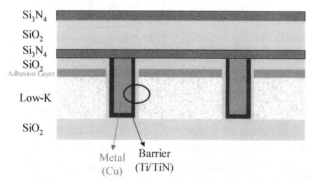

Fig. 14. The Ti/TiN barrier layer is itself protected by an SiO₂ layer. The Si₃N₄ layers serve as the etch stop. The need for an adhesion layer between SiO₂ and low-k dielectric depends on which is deposited first.

chemical stability, adhesion, and gap fill capabilities.

Low Dielectric Constant Materials

- SiOF: fluorinated silica
- SiOH: hydrogenated silica
- Porous SiO_2
- Organic polymers
- Fluorinated polyimide
- Fluorinated arylene ether
- Parylene
- PTFE (Teflon)
- Amorphous fluorinated carbon (Figs. 12 and 13)
- Air gap
- Hybrid inorganic-organic material: F-polyimide + SiO_2

Desirable characteristics

- Dielectric constant: k ~ 1 – 3
- Good thermal stability, low expansion
- Minimal moisture uptake
- Good mechanical strength
- Electrical leakage/breakdown similar to SiO_2
- Less film stress
- Good adhesion
- Less capping material
- CMP compatible
- Etching selectivity to nitrides/oxides/oxynitrides
- O_2 ashing compatible

3. Barrier materials

Titanium, tantalum, and tungsten and/or their nitrides with low resistance are used as diffusion barriers and adhesion promoters between copper and dielectric. The integration of barrier materials with metal and ILD is critically important in microelectronics interconnection, especially the chemical and thermal stabilities at the metal/barrier and barrier/ILD interfaces (Fig. 14). Atomic layer deposition has been widely studied recently to allow the deposition of these materials over high aspect ratio features with greater conformality.

III. PATTERNING NANOMETER FEATURES

1. E-beam

To generate finer features than uv light can define, electrons generated from a thermionic or field-emission source can be used to "write" photoresist

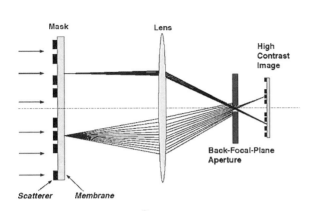

Fig. 15. The SCALPEL$^{\circledR}$ e-beam lithography system.

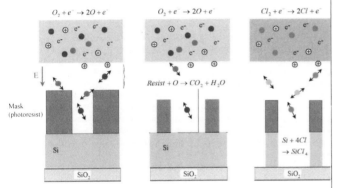

Fig. 16. By etching the photoresist, it can be made narrower to generate finer features.

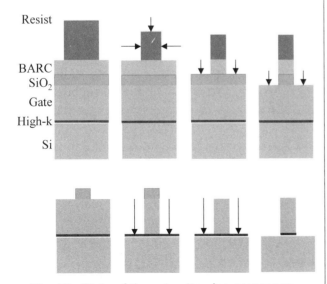

Fig. 17. State-of-the-art patterning processes.

directly. Electron scattering and the resultant "proximity" effect are the major challenges in e-beam lithography. The SCALPEL$^{\circledR}$ (Scattering with Angular Limitation in Projection Electron-beam Lithography) system developed by Bell Labs, Lucent Technologies, uses high-energy electrons, projected through a photomask, to create integrated circuit features just 30-80 nm wide, overcoming many of the limitations faced in the current optical lithography systems due to the available wavelengths of light. In Fig. 15, an electron beam enters from the left and impinges on a mask thick enough to scatter them but not to stop them. Electrons passing through the open areas suffer minimal scattering and diffraction. An electromagnetic lens refocuses the electrons onto the photoresist through a back-plane aperture, placed so that most of the unscattered electrons but few of the scattered ones will be transmitted.

A group of semiconductor device and equipment manufacturers has recently announced a joint agreement aimed at accelerating the development of SCALPEL$^{\circledR}$ technology into a production lithography solution for building future generations of integrated circuits.

2. Resist trimming

Another alternative to generate finer features is to "trim" the photolithographically patterned features by an oxygen plasma etching process, as shown in Fig. 16. In a real process concerning patterning the gate stack, several etching processes have to be used because the photoresist is deposited on a layer of bottom anti-reflective coating (BARC) on a layer of SiO_2, which is used as the hard mask in patterning the gate polysilicon, as shown in Figure 17. It is then necessary to use a series of plasma etching processes to (1) trim the photoresist, (2) etch the BARC, (3) pattern the hard mask, (4) ash the photoresist and BARC, (5) etch the gate, (6) remove the hard mask, and (7) etch the gate dielectric. Each process requires a different etch chemistry and an in-depth understanding of the surface chemistry to achieve high etch rate, greater anisotropy, high selectivity, and less damage. This demonstrates the complexity of the process of fabricating nanometer-scale transistors, and the challenge is to simplify and automate the process for speed, reliability, and reproducibility.

IV. DEEP REACTIVE ETCH FOR MEMS

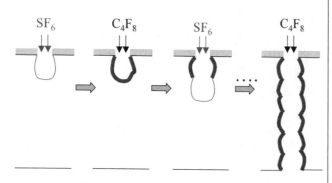

Fig. 18. Schematic of the BOSCH process.

Fig. 19. High aspect ratio features patterned with the BOSCH process.

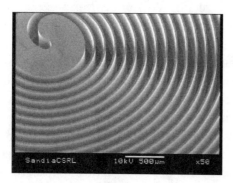

Fig. 20. MEMS devices (a micro-GC) made by the BOSCH process (Sandia National Lab).

Plasma etching is also needed in generating novel MEMS (MicroElectroMechanical Systems) structures, especially for high aspect ratio features. Because the dimensions of MEMS devices are much larger than in microelectronics, much faster etch rates (100 μm/h) is needed, but higher tolerance and some isotropy are allowed. To meet these requirements, the BOSCH process is typically used (Figs. 18 and 19).

The key element in the BOSCH process is to alternate etching and passivation processes to allow the generation of high aspect ratio features with high throughput. SF_6 is typically used to etch silicon to obtain high etch rate, and C_4F_8 is used to coat the undercut area with a passivation layer. By iterating these steps, high aspect ratio features can be generated, typically with scalloped ridges inside the feature. Figure 20 shows a spiraled gas chromatography column (GC) that is etched through a silicon wafer using the BOSCH process. Once this spiraled column is coated with the appropriate stationary phase material and encapsulated, a GC is made and can be used for chemical separation prior to detection.

V. PLASMA-INDUCED DAMAGE

The delicate circuits on a chip are easily damaged during plasma processing by bombardment by energetic ions or electrons, or even by ultraviolet radiation. Most problems are caused by the thin gate insulator, usually SiO_2. Suppose the gate is charged to 0.4V by the plasma while the other side of the oxide is at ground potential. If the oxide is 40Å thick, the E-field across it is 1 MV/cm! High E-fields cause a quantum-mechanical current, called the *Fowler-Nordheim* current, to tunnel through the dielectric. If the oxide layer is much thinner than this, not much damage is incurred, but the layer is no longer a good insulator. If it is thicker than this, electrons driven through the layer will damage the dielectric by creating defects in the lattice structure, changing the oxide's characteristics, such as its capacitance or breakdown voltage. Initially, plasma potential uniformities or $\mathbf{E} \times \mathbf{B}$ drifts due to imposed B-fields were blamed for charging damage, but recently attention has been focused on the *electron shading* mechanism proposed by Hashimoto [Jpn. J. Appl. Phys. **33**, 6013 (1994)].

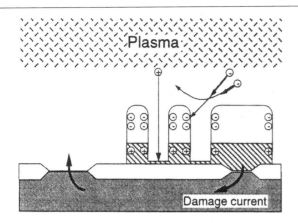

Fig. 21. Hashimoto's diagram of the electron shading mechanism.

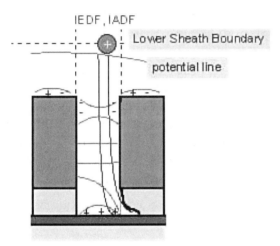

Fig. 22. Monte-Carlo simulation of ion orbits affected by photoresist charging. In this case, curved orbits can lead to notching of the trench bottom [Hwang and Giapis, JVSTB **15**, 70 (1997)].

In electron shading (Fig. 21), electrons impinge on the insulating photoresist and charge it negatively. The resultant negative space potential prevents further electrons from entering the trench. Only ions reach the trench bottom, causing it to charge up positively. If this potential is connected to a gate electrode, current will be driven through the oxide. The current flows back to ground elsewhere on the chip to complete the circuit. Note that if the gate is connected to ground during the etch, no charge buildup can occur. However, when the etching is complete and the gate is isolated, then it can become charged. Thus, most of the damage occurs just before or during the *overetch* period. The amount of charge available to damage an oxide depends on other charge-collecting areas that are connected to the gate. The *antenna ratio* is ratio of this total area to the oxide area and can be of the order of 10^4. The increase in damage with antenna ratio is well documented. Special diagnostic wafer have been developed to measure the probability of damage in various parts of the wafer.

There has been evidence that low T_e can reduce damage, presumably because all plasma voltages decrease with T_e. Pulsed discharges have been studied for this purpose. In the afterglow plasma after RF turnoff, T_e decays faster than does n, so that there is a period when etching can be done with a low-T_e plasma. Mechanical filters for fast electrons have been proposed, as well as the introduction of gases that absorb them. On the other hand, RIE discharges, which normally have higher T_e than ICPs, have been found to cause less damage. Though there have been computer simulations (Fig. 22), Hashimoto's scenario has not been verified directly, and the primary mechanism for oxide damage is not well enough understood that curative measures can be taken in a predictive fashion.

VI. SPECIES CONTROL IN PLASMA REACTORS

Though plasma reactors are available which can produce uniformity in density and temperature over a 300-mm substrate, it is more difficult to control the uniformity of various chemical species impinging on the wafer. There are two problems: uniformity of the neutral gas, and the distribution of molecular species.

In processes that consume large quantities of

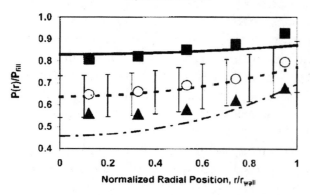

Fig. 23. Neutral depletion of an initially uniform filling of 10-mTorr argon gas by the plasma in a helicon source. The data are at P_{rf} = 1, 2, and 3 kW (top to bottom), and the lines are theoretical [G.R. Tynan, J. Appl. Phys. **86**, 5356 (1999)].

the injected gas, such as deposition of amorphous silicon onto glass substrates for flat-panel displays, large-area, small-gap RIE discharges can be used. The gas is injected uniformly through hundreds of small holes in one of the capacitor plates. In low-pressure etching with ICPs, however, neutrals cannot easily be injected into the interior of the plasma. A dense plasma will ionize neutrals injected from a showerhead ring around the periphery, leaving the center depleted of neutrals. This has been observed by Tynan (Fig. 23).

A more difficult problem is to control, for instance in an argon–fluorocarbon discharge, the relative concentrations of CF_4, CF_3, CF_2, CF, F, etc. at the wafer level. Some degree of control can be obtained by creating an argon plasma in the upper part of the chamber and then injecting the reactant gas from a showerhead lower down, where T_e has cooled to a suitable level. In traversing the plasma toward the center, however, neutrals will be dissociated and ionized progressively, and the relative concentrations will vary with radius. Plasmas with electronegative gases pose further problems. Control of the physical properties of RF plasma may be at hand, but control of the chemical properties, perhaps by controlling the EEDF, is a subject for further development.